Stars & Planets
atlas

Ian Ridpath

Facts On File, Inc.

Endpapers *Cosmic clouds – an area of glowing hydrogen gas on the borders of Cepheus and Cassiopeia, captured with a long-exposure photograph.*

First published in Great Britain in 1992 by George Philip Limited,
a division of Octopus Publishing Group Limited,
2–4 Heron Quays, London E14 4JP

Copyright © 1992, 1997, 2001 George Philip Limited
Second edition 1997
Third edition 2001

First published in the United States of America by Facts On File, Inc.

Facts On File, Inc.
132 West 31st Street
17th Floor
New York, NY10001

Facts On File books are available at special discounts when purchased in
bulk quantities for businesses, associations, institutions or sales promotions.

Please call our Special Sales Department in New York
at (212) 967–8800 or (800) 322–8755.

You can find Facts On File on the World Wide Web at http://www.factsonfile.com

Library of Congress Cataloging-in-Publication Data available

ISBN 0–8160–4800–2

Printing (last digit): 10 9 8 7 6 5 4 3 2 1

Printed in Hong Kong

Contents

The Solar System

Our home in space is the planet Earth. It is one of nine planets that go around the Sun, which is a star. The Earth is the third planet from the Sun. Each planet is a world in its own right, with its own surface features and conditions. For example, the biggest planets – Jupiter, Saturn, Uranus, and Neptune – are not solid, but are balls of liquid and gas that would be impossible to land a spaceship on.

The path of one body around another in space is called its *orbit*, and the time taken for the Earth to go all the way around its orbit is one year. The closest planets to the Sun move around their orbits much more quickly than those farther away. For example Mercury, the innermost planet, speeds around the Sun once every 3 months, but distant Pluto takes nearly 250 of our years to complete its "year." You would never live long enough to reach your first birthday on Pluto! Together, the Sun and all the objects that orbit it make up the Solar System.

As well as moving along their orbits, the planets spin. The Earth spins once every 24 hours, which we call a day. Some other planets spin more quickly than the Earth, others more slowly. The Earth has a smaller companion, the Moon, going around it. Most planets have at least one moon. Only Mercury and Venus, the two planets closest to the Sun, are moonless.

The orbits of the planets around the Sun are almost circular, except for Pluto's. For most of the time Pluto is the most distant planet from the Sun, but part of its orbit brings it closer to the Sun than its neighbor planet Neptune, as was the case between 1979 and 1999. Fortunately, Pluto and Neptune will never collide because their two orbits are tilted in such a way that they do not cross.

In addition to the nine planets, there is a lot of debris in the Solar System. Look in the picture for a band of rubble called the *asteroids* between the planets Mars and Jupiter. Other small, frozen bodies called *comets* move on highly elongated paths. When they come close to the Sun they heat up and grow long, flowing tails.

asteroids

▼ *Here the distances of the planets from the Sun are shown to scale. Look how "bunched up" the inner planets are, compared with the outer planets. But the distances between them are still enormous.*

Neptune

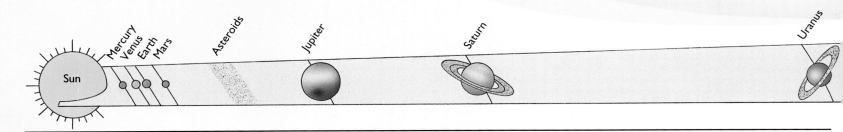

Sun Mercury Venus Earth Mars Asteroids Jupiter Saturn Uranus

Other solar systems?

In recent years, astronomers have begun to detect planets around some nearby stars. All the planets found so far are large, like Jupiter, but smaller planets like Earth may also exist. This artist's impression shows an imaginary view above the moon of a giant planet in another solar system. Some of the planets around other stars may have living things on them.

comet

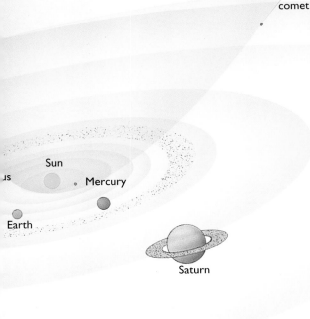

Sun

Mercury

Earth

Saturn

Uranus

▲ The Sun and its family of nine planets, from Mercury to Pluto. Except for Mercury and Venus, all the planets have moons going around them, and the biggest planets have rings too. This picture is not to scale. To get an idea of how far the planets are from one another, look at the diagram below.

Until about 400 years ago, people thought that the Earth was the center of the Universe and that everything else went around it. In 1543 Nicolaus Copernicus, a Polish astronomer and churchman, suggested that the Earth was a planet that orbited the Sun. Later astronomers proved him right. This stamp shows Copernicus and his diagram of the Solar System with the Sun at the center.

Pluto

Neptune

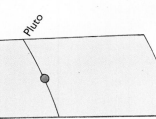

Pluto

The Sun

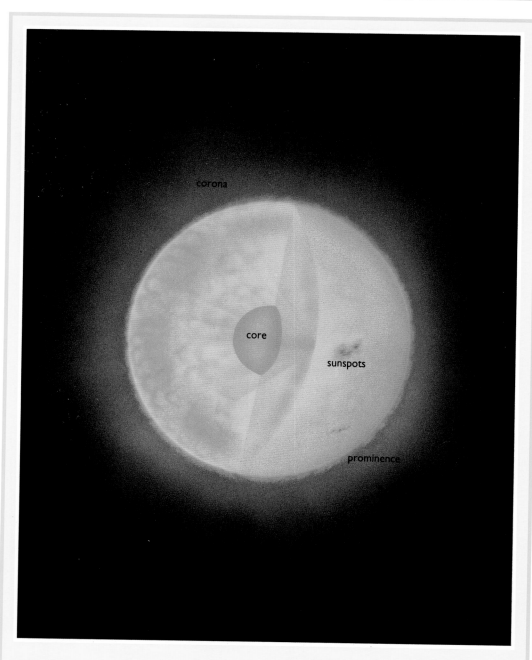

Every morning the Sun rises in the east, bringing a new day. Ancient peoples regarded the Sun as a god, but we know that it is actually a star, a blazing ball of gas that gives out light and heat, which planets do not. Without the light and heat from the Sun, there would be no life on Earth.

The Sun is huge: 1.4 million kilometers (865,000 miles) in diameter. Over 100 Earths would be needed to stretch in a line from one side of the Sun to the other. Yet, as stars go, the Sun is only average in size and brightness. It looks much bigger and brighter than the stars at night because it is so much closer. Even so, the Sun is still 150 million kilometers (93 million miles) from the Earth. This is just as well for us because its surface is a sizzling 5500°C (9900°F). We would be roasted if we went too close. At the Sun's core, the temperature increases to an incredible 15 million degrees.

What keeps the Sun glowing? Until about 1940, this was a mystery. The Sun cannot burn like a lump of coal, for there is no air in space to feed the flames, and in any case it would have burned out long ago. One clue is that the Sun is composed entirely of gas — mostly hydrogen, the lightest gas known, plus some helium.

Structure of the Sun

Energy is released at the center of the Sun. It makes its way to the surface, where it is radiated into space as heat and light. The surface of the Sun is called the *photosphere* ("sphere of light"). On it are dark markings called *sunspots*. Above it is a thinner layer of gas called the *chromosphere* ("color sphere"). Loops of gas called *prominences* reach into space from the Sun's surface. Around the Sun is a faint halo of gas called the *corona*, which can be seen only at eclipses. A stream of atomic particles from the Sun called the *solar wind* blows outwards past the planets, including the Earth.

SUN DATA

Diameter	865,000 mi
Mass:	333,000 × Earth
Volume:	1.3 million × Earth
Average density:	1.4 × water
Time to spin on axis:	25.4 days (average)
Distance from Earth:	93 million mi

During the 1930s, scientists came to realize that the Sun is an enormous nuclear reactor. Inside the Sun, hydrogen is turned into helium by a process known as *fusion*. What happens is that atoms of hydrogen are crushed together (fused) by the enormous temperatures and pressures at the Sun's core, creating atoms of helium. In the process, energy is given out and this energy makes the Sun hot. Every second, 600 million tons of hydrogen is turned into helium. But the Sun is so big that it has enough hydrogen left to carry on burning for thousands of millions of years – and it is already 4.6 billion years old! There is no danger of the Sun going out for a long while yet.

◀ *Sunset over the Adriatic Sea, photographed from the coast of Croatia. At sunrise and sunset, the Earth's atmosphere makes the Sun look redder.*

proton   proton

proton   proton

energy   energy

neutron   proton

proton   neutron

energy   energy

Energy inside the Sun

In the fusion process that powers the Sun and other stars, four atoms of hydrogen are crushed together to make one atom of helium. Energy is released in the reaction. Scientists are working to harness the energy of fusion for power stations on Earth. But it is a difficult job because the process needs extremely high temperatures and pressures to work.

▶ *Inside the Joint European Torus, an experimental reactor in which scientists have recreated the same fusion reactions that power the Sun. This is the first step towards fusion power stations here on Earth.*

Features of the Sun

The Sun's boiling surface is marked by dark patches called sunspots. When scientists began to study the Sun regularly through telescopes, they found that the number of sunspots rises and falls in a cycle lasting about 11 years. Even now, no one fully understands the reason for these changes. But we do know that sunspots are actually cooler areas on the Sun's surface. By everyday standards they are still quite hot – even the coolest part of a sunspot is at a temperature of about 4000°C (7200°F). But they appear dark by contrast with the brighter surface around them, which is 1500°C (2700°F) hotter.

A typical sunspot is many times larger than the Earth. Spots often occur in groups, and some groups can stretch for 100,000 kilometers (60,000 miles) or more, a quarter of the distance from the Earth to the Moon. Sunspots live for anything from a few days to a few weeks before fading out. As the Sun rotates, the spots pass across the Sun's face from west to east. By watching sunspots, astronomers can see that the Sun rotates fastest at the equator – once every three and a half weeks – but more slowly towards the poles.

▼ *A sunspot group photographed near the edge of the Sun's disk. Many flares erupted near these particular spots, producing aurorae in the Earth's atmosphere.*

From time to time huge explosions called *flares* occur near sunspots. In a few minutes, a flare releases as much energy as millions of hydrogen bombs, and fires atomic particles into space at high speeds. These particles reach the Earth after a day or two, where they bombard the upper atmosphere to cause colorful nighttime glows known as *aurorae*.

Clouds known as *prominences* lie above the Sun's surface. They extend tens of thousands of kilometers into space and can be seen around the Sun's rim at a total eclipse. Prominences are often shaped like arches, because they follow looping lines of magnetism. Some prominences, though, consist of material being sprayed out by flares, or falling back after a flare has died away.

▼ *Large amounts of energy were released by this flare, which burst out two days after the photograph on the left was taken. It was one of the largest and brightest ever seen.*

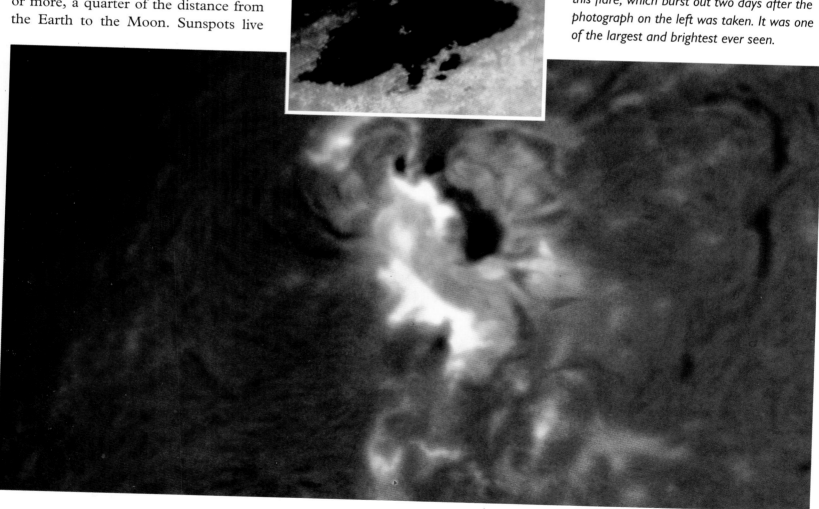

▼ A flame-like prominence shoots out from a flare – 10 billion tons of hydrogen gas moving at 1.5 million kilometers (1 million miles) per hour. Prominences like this last no longer than a few hours.

▶ Several large sunspots speckle the face of the Sun. All the photographs of the Sun on these two pages were taken in 1989, a year when the sunspot cycle was at a peak and there was a lot of activity on the Sun.

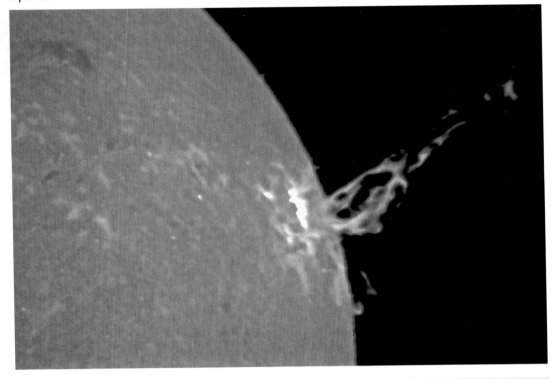

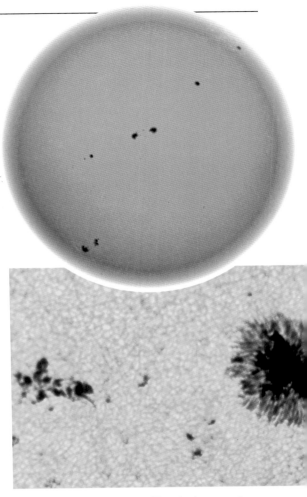

▲ A group of sunspots. The dark central part of a sunspot is called the umbra, and the lighter outer part is called the penumbra. These two regions can be seen in the large spot on the right. The speckled appearance is termed granulation, and is caused by bubbles of hot gas rising to the Sun's surface.

Aurora

An aurora is a colorful glow high in the Earth's atmosphere caused by atomic particles from the Sun. This aurora was photographed from space by astronauts aboard the Space Shuttle *Discovery* in May 1991. It rises to 250 kilometers (150 miles) above the Earth. Aurorae are most often seen near the Earth's magnetic poles but sometimes, following a big flare or coronal mass ejection on the Sun, they can be seen as far south as the Mediterranean or the southern United States.

A space probe called Ulysses, launched in 1990 by the European Space Agency to study the Sun, is shown on this stamp from Hungary. Ulysses flew over the north and south poles of the Sun, studying it from directions not seen from Earth.

Planet Earth

From out in space, our home planet looks like a blue-and-white marble. The blue is the oceans and the white parts are clouds in the planet's dense atmosphere. Three-quarters of our planet is covered with water, and the Earth is the only planet in the Solar System with lots of it.

Water and an atmosphere are two reasons why there is life on the Earth. Another important reason is that the Earth's distance from the Sun, about 150 million kilometers (93 million miles), is just right. If it were closer to the Sun it would be too hot for life, like Venus. Farther away, and it would be too cold, like Mars.

The Earth's crust is cracked into pieces, like a damaged eggshell. These pieces are moving slowly, and volcanoes erupt along the cracks. In the past, the Earth had many more volcanoes than today. Most of our atmosphere and water is thought to have been given out by ancient volcanoes, although some may also have come from comets that hit the Earth.

The atmosphere is mostly nitrogen, with about 20 percent oxygen, which we take in when we breathe. The oxygen is released by plants that break down carbon dioxide. A form of oxygen called *ozone* protects the Earth from the Sun's dangerous ultraviolet light. But satellites in orbit around the Earth have found evidence that this layer is

thinning out because of man-made chemicals being released into the atmosphere.

Satellites keep watch on the Earth's atmosphere and oceans, measuring temperatures, winds speeds, moisture content, and ocean currents. Scientists use such measurements to predict the weather and to see if the climate is changing. Other satellites survey the Earth's land surface and the things that live and grow on it.

The Earth has one moon, which is one-quarter its size. This is larger in comparison with the planet than any moon except the one orbiting the tiny planet Pluto. In a way, therefore, the Earth and Moon are like a double planet.

▲ Earthrise over the Moon. If you lived on the Moon, you would see the Earth pass through the same cycle of phases that we see the Moon pass through from Earth.

▼ Hurricane Fefa swirling over the western Pacific Ocean in August 1991. Photographs like this one taken from space are important to scientists studying the Earth's weather.

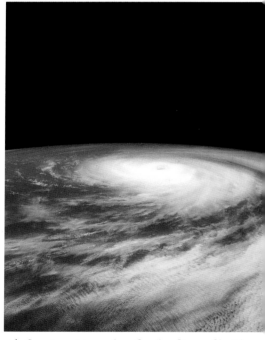

◀ Sunrise approaches for the Space Shuttle Atlantis *as it orbits the Earth.*

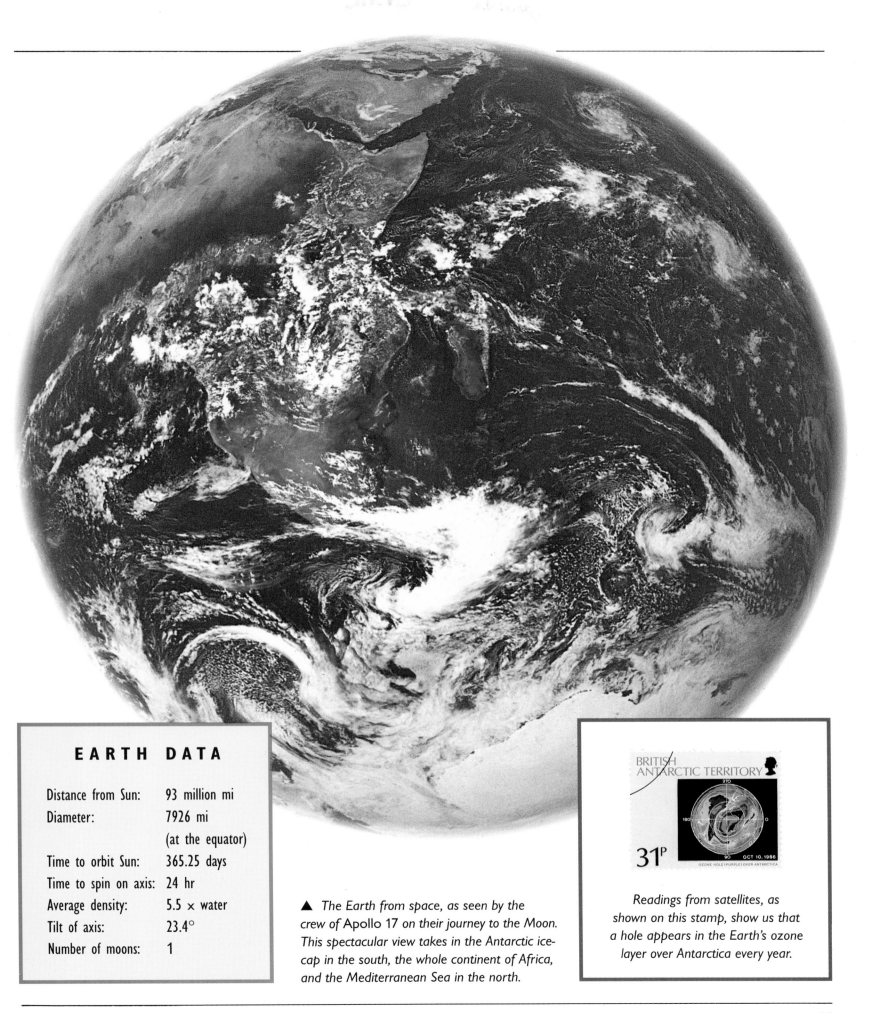

EARTH DATA

Distance from Sun:	93 million mi
Diameter:	7926 mi
	(at the equator)
Time to orbit Sun:	365.25 days
Time to spin on axis:	24 hr
Average density:	5.5 × water
Tilt of axis:	23.4°
Number of moons:	1

▲ The Earth from space, as seen by the crew of Apollo 17 on their journey to the Moon. This spectacular view takes in the Antarctic ice-cap in the south, the whole continent of Africa, and the Mediterranean Sea in the north.

BRITISH
ANTARCTIC TERRITORY

270
180   O
90   OCT 10, 1986
OZONE HOLE (PURPLE) OVER ANTARCTICA

31ᴾ

Readings from satellites, as shown on this stamp, show us that a hole appears in the Earth's ozone layer over Antarctica every year.

Time and the Earth

We do not normally notice it, but the Earth is continuously on the move. Every year it goes once around the Sun, a journey of nearly 1 billion kilometers (600 million miles), at a speed of over 100,000 kilometers (60,000 miles) per hour. The journey takes just over 365 days, and our calendar is based on it.

As the Earth moves around its orbit, the Sun appears to change position against the background stars. The Sun traces out a yearly path around the sky, called the *ecliptic*, which is marked on the star maps on pages 54–57. The ecliptic is tilted by about 23.5° with respect to the equator, because of the tilt of the Earth's axis. In prehistoric times, people built stone circles such as the one at Stonehenge in England to track the changing position of the Sun throughout the year. Such stone circles were the first calendars.

As well as moving along its orbit, the Earth also spins on its axis, turning once every day. This makes the Sun rise and set, and the stars cross the sky at night. Our time system is based on the daily movement of the stars. Astronomers still make accurate observations of the movement of the stars, to keep our clocks in step with the rotation of the Earth.

The Earth's day, the time it takes to spin once on its axis, is just under 24 hours. But the tides, described on page 21, are very gradually slowing down this spin and making the day longer – 200 million years from now the day will be 25 hours long.

▲ *As the Earth turns, the stars appear to move across the sky. In this time-exposure photograph, the stars are drawn out into curved trails. The bright, short trail at the upper right is Polaris, the north pole star. Polaris does not lie exactly at the pole of the sky – if it did, it would not leave a trail. In the foreground is the dome of the William Herschel Telescope on La Palma, in the Canary Islands.*

▶ *This amazing multiple-exposure photograph shows the changing position of the Sun at 8.30 a.m. over an entire year. The Sun is highest in summer (top left) and lowest in winter. The three bright streaks are the rising Sun. The figure eight shape occurs because the Earth's axis is tilted and its orbit around the Sun is not quite circular.*

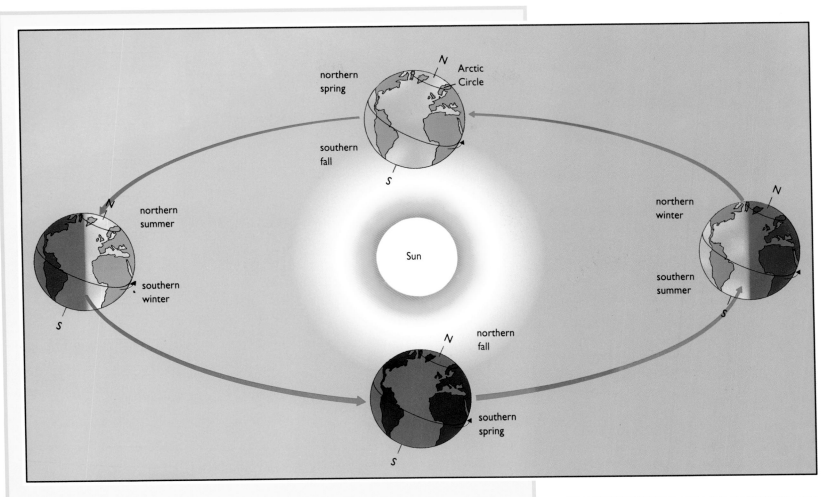

The seasons

There are seasons because the Earth's axis is tilted. At the *summer solstice*, the Earth's north pole leans towards the Sun and days in the northern hemisphere are longest. Within the *Arctic Circle*, the Sun never sets. Six months later, at the *winter solstice*, the situation is reversed. In between, at the spring and autumn (fall) *equinoxes*, the Sun lies on the celestial equator and day and night are roughly equal the world over.

▲ *This man is fishing in Finnmark, in northern Norway. Finnmark is inside the Arctic Circle, so even though the photograph was taken at midnight, the Sun is still shining because it is summer. In the winter, the Sun never rises.*

Greenwich Mean Time (GMT) is used as a time standard around the world. It is time as measured on the Greenwich meridian, the 0° line of longitude, which passes through the Greenwich Observatory in London. It is shown on this stamp, which commemorates the 100th anniversary of the Greenwich meridian.

▶ *Midsummer sunrise at Stonehenge, on Salisbury Plain in England. This stone structure dates back 4000 years, and was used to observe the times and positions of sunrise throughout the year. Stone circles like Stonehenge were the first observatories.*

The Earth at night

Surprisingly, signs of life on Earth are easier to see at night than by day. This remarkable view of the Earth at night was put together from about 40 photographs taken by weather satellites.

The bright areas are all man-made lights, mostly from streets and buildings but with some fires as well, so they outline the developed areas of the world.

Look, for example, at the north-eastern United States, northern Europe, and Japan. In North America, Europe and Japan, one-quarter of the world's

people use about three-quarters of the world's electricity. All of the light shown on this picture is wasted – it is shining upwards into space. Astronomers call this *light pollution*, and it makes the night sky brighter so that stars are difficult to see.

Large blobs of light in the Middle

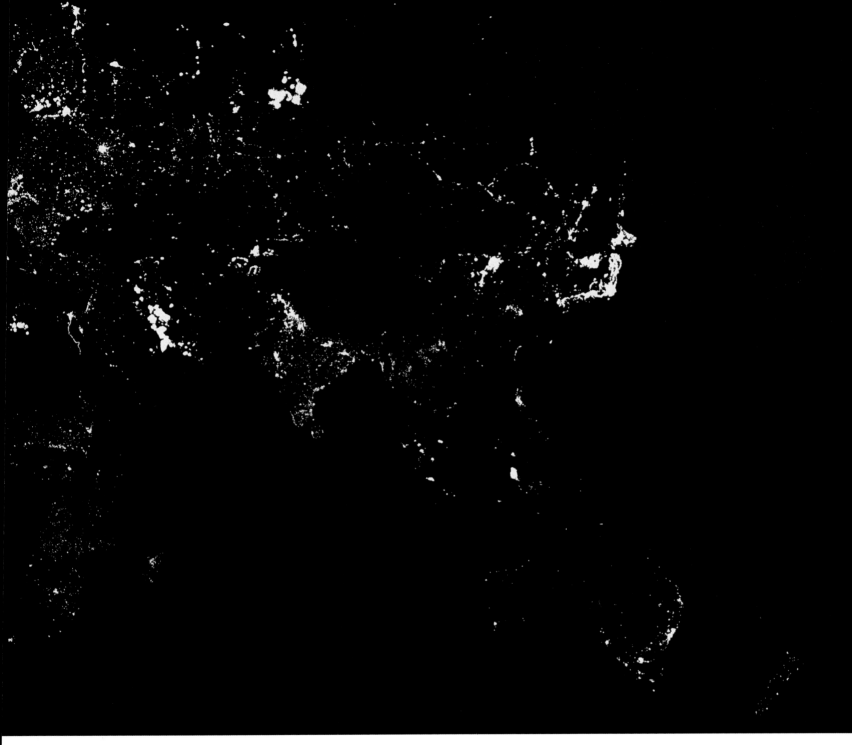

East, central Asia, and North Africa are from gas being burnt in oil fields. Another large blob of light, this time near Japan, comes from fishing boats shining bright lights to attract squid. Small specks of light across Africa and Southeast Asia are fires in grassland and forests. In parts of the world, people use fire to clear areas for cultivation. This reduces the Earth's forests and adds carbon dioxide gas to the atmosphere, which may affect the climate.

Several interesting details can be picked out, such as the spoke-like pattern of roads around Moscow and other large cities, a chain of lights along the Trans-Siberian railway, and the River Nile. Some coastlines, such as those of Florida and Spain, are outlined by lights. Puerto Rico shines brightly in the Caribbean Sea. France wastes much less light than its neighbors in Europe.

The only natural source of light in this picture is an aurora over Greenland.

The Moon

▲ *Three photographs of the Moon, taken at crescent phase, first quarter, and full Moon. Some features change their appearance as the phase changes. For example, the line of craters running south from the middle of the Moon's disk is quite hard to pick out at full Moon, but it stands out clearly at first quarter.*

The Moon is the Earth's companion in space, lying roughly 380,000 kilometers (240,000 miles) away, a distance that would take two or three days to travel in a spacecraft. It is a ball of rock about one-quarter of the size of our planet, with no air, water, or life. It is the only other body in space on which human beings have ever set foot. American astronauts walked on the Moon during the Apollo space missions, from 1969 to 1972.

The Moon moves around the Earth, so it rises and sets at different times each night. As it orbits the Earth, the Moon seems to change in shape – sometimes it is a crescent (curved), other times it is a half-circle, and other times it is a full circle. These changes are known as the Moon's *phases*, and

the diagram explains how they occur. A complete cycle of phases lasts 29 days and is the origin of our month.

Even a quick glance at the Moon shows that it has dark markings which form a pattern like a face, popularly called the "man in the Moon." Every time we look at the Moon we see the same markings, which means that the Moon keeps the same side turned towards us. There is a very good reason for this. The Earth's pull of gravity has "locked" the Moon so that it turns on its own axis in the same time as it takes to orbit the Earth.

MOON DATA

Diameter:	2160 mi
Distance from Earth:	238,900 mi
Time to orbit Earth:	27.32 days
Time to spin on axis:	27.32 days
Mass:	0.0123 × Earth
Volume:	0.02 × Earth
Average density:	3.3 × water

▶ *As the Moon orbits the Earth, we see different amounts of its sunlit side. When the Moon lies between the Earth and Sun we do not see it at all; this is known as new Moon. Then it becomes a young crescent and appears low in the western sky in the evening. A few days later it becomes half-illuminated, known as first quarter. The phase between half and full Moon is called gibbous. After full Moon there are the same phases but in reverse, ending with a crescent Moon rising in the morning sky shortly before the Sun.*

▼ *One of the most prominent lunar craters is called Copernicus, and this photograph of it was taken from the Apollo 12 spacecraft. It is 93 kilometers (58 miles) across.*

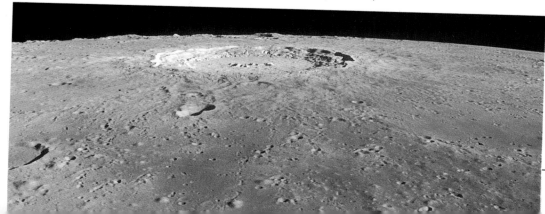

Where did the Moon come from?

Astronomers are still not sure how the Moon was born. One old theory said it was a piece of the Earth that broke off long ago, when our planet was hot and spinning quickly. Another idea was that it was once a separate body that was captured by the Earth's gravity when it passed too close. In a third theory, the Moon grew from material left over from the birth of our planet. But there are problems with all three theories.

A more recent theory says that the Moon originated in a huge collision between the Earth and another body. The other body was destroyed in the impact, but bits from it and the Earth were blasted into space. They created a ring around the Earth, and the debris later gathered together to form the Moon.

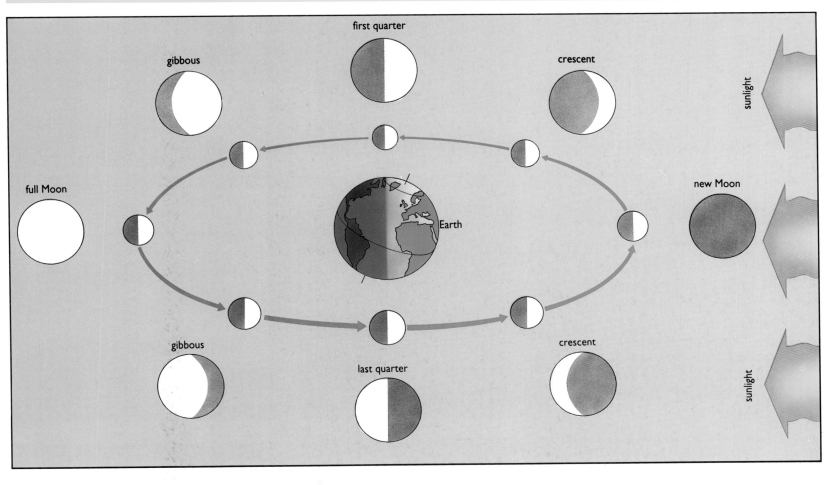

first quarter

gibbous

crescent

sunlight

full Moon

Earth

new Moon

gibbous

crescent

last quarter

sunlight

The Moon's features

A simple pair of binoculars will show an amazing amount of detail on the Moon. There are dark, smooth lowlands and bright, mountainous areas dotted with craters of all sizes. The dark lowlands are called "seas," although there is no water on the Moon, nor has there ever been. The seas have fanciful names such as Mare Nubium (Sea of Clouds), Oceanus Procellarum (Ocean of Storms) and Mare Tranquillitatis (Sea of Tranquillity).

The craters were formed by meteorites and comets hitting the Moon long ago. The largest craters are over 100 kilometers (60 miles) across, large enough to swallow a city. They are named after scientists and other famous people. Asteroids also hit the Moon, digging out huge basins. These basins were later flooded by volcanic lava from inside the Moon, forming the seas.

The Moon's surface is incredibly ancient. Rocks from the Moon's seas,

The far side

Because the Moon always keeps one face turned towards us, no one had ever seen the far side of the Moon until space probes flew around it. They showed that most of the Moon's far side is covered with heavily cratered highlands, with few of the dark seas that are common on the Earth-facing side. The reason for this is that the Moon's crust is thicker on the far side, so that volcanic lava could not easily flow onto the surface there.

▶ *Two photographs of the Moon, at first quarter and last quarter, with some of the main features labeled. Capital letters are used for the names of lowland areas. The first letter of each crater name is just to the right of the crater itself. To match these pictures with the view through a telescope, turn the book upside-down. Binoculars will show the Moon the "right" way up.*

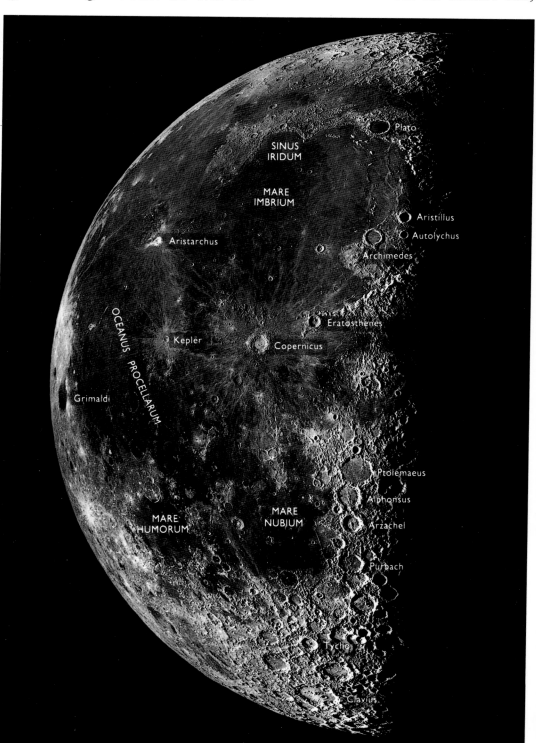

brought back by Apollo astronauts, are over 3 billion years old. The oldest rocks, from the lunar highlands, are 4.5 billion years old, and so date back to the birth of the Moon.

The Moon has changed very little since the seas were formed, except where a few meteorites have produced new craters. You can easily spot the youngest craters, because they have long white rays stretching away from them. The rays are particularly bright at full Moon, and consist of rock thrown out by the impact that made the crater. Rays from one bright crater, Tycho, stretch over most of the Moon's near side. Tycho is about 100 million years old, one of the youngest of the large craters.

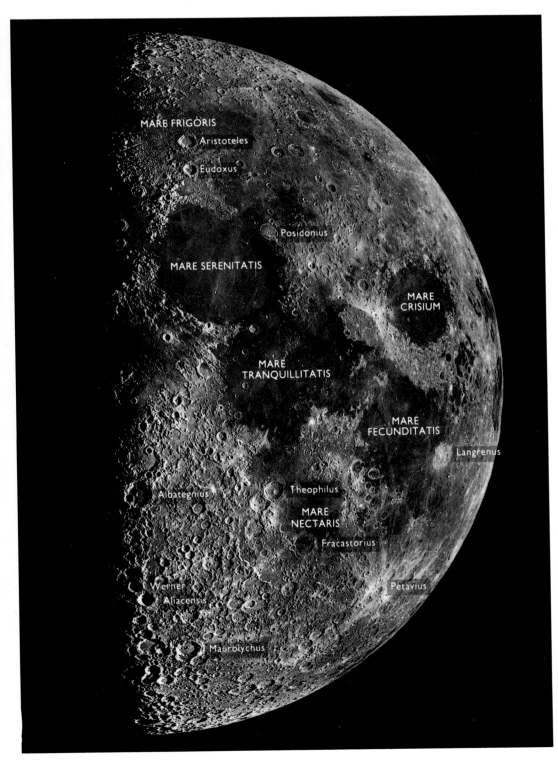

Man on the Moon

▲ Astronaut Charles Conrad Jr., commander of the Apollo 12 mission, at work on the Moon's surface in November 1969. This was the second manned lunar landing.

Neil Armstrong and Edwin Aldrin were the first humans to land on the Moon. They touched down on the Sea of Tranquillity on July 20, 1969, during the Apollo 11 space mission. As Armstrong stepped onto the Moon, he said: "That's one small step for a man, one giant leap for mankind." Armstrong and Aldrin walked on the Moon for two hours, collecting 20 kilograms (48 lb) of rocks. Later Apollo missions took an electric Moon car which astronauts drove over the surface. In all, twelve Apollo astronauts walked on the Moon. The last mission, Apollo 17, was in December 1972.

Neil Armstrong, the first human to set foot on the Moon, is pictured on this stamp.

Eclipses and tides

The Earth and Moon each cast a shadow into space, and an eclipse happens when one of them goes into the other's shadow. An eclipse of the Sun is when the Moon moves in front of the Sun, and its shadow falls on the Earth. An eclipse of the Moon is when the Earth's shadow falls on the Moon. There are at least two eclipses of the Sun every year and usually one or two of the Moon, but they cannot all be seen from one place on Earth.

Most spectacular are *total eclipses* of the Sun, when the Moon completely blots out the Sun's bright face and turns day into night for several minutes.

◄ *The Sun's beautiful corona, a halo of gas, seen from Kenya during a total eclipse on February 16, 1980.*

▼ *How eclipses happen. When the Moon's shadow falls on the Earth (left), the Sun is eclipsed. The dark, inner part of the shadow is called the* umbra. *Where the path of the umbra passes across the Earth, people see a total eclipse. In the lighter, outer part of the shadow, the* penumbra, *the eclipse is partial. When the Earth's shadow falls on the Moon (right), the Moon is eclipsed.*

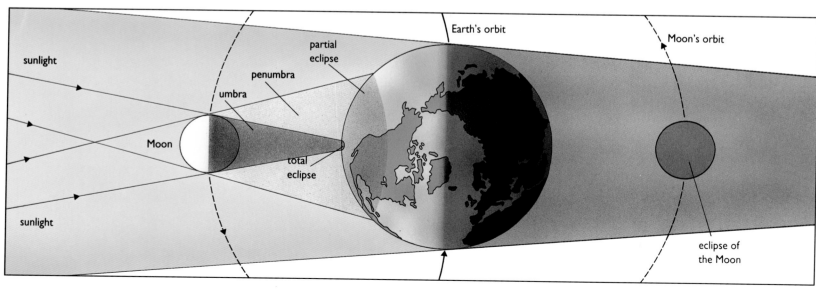

▲ *An eclipse of the Moon, in August 1989.*

During a total solar eclipse, the faint halo of gas around the Sun, its corona, comes into view. Astronomers often travel the world to see a total eclipse of the Sun. A *partial eclipse*, in which the Sun is only partly covered by the Moon, is seen over a much wider area of Earth than the total eclipse.

Total solar eclipses are the result of a remarkable coincidence: the Moon and Sun appear almost exactly the same size in the sky. If the Moon were smaller, or farther away, it would not cover all of the Sun. In fact, the Moon's distance from Earth changes slightly, because its orbit is not exactly circular. If a solar eclipse happens when the Moon is at its most distant from us, the eclipse is not total. Instead, a ring of sunlight is left around the Moon at mid-eclipse. This is known as an *annular eclipse*, from a Latin word meaning "ring."

Total eclipses of the Sun can last up to seven and a half minutes, but most are much shorter. Total eclipses of the Moon can last for over an hour, and can be seen from anywhere the Moon is above the horizon. However, the Moon does not disappear, even when it lies entirely within the Earth's shadow and is thus totally eclipsed. Usually, it turns a dark red color. This is because some of the Sun's light is bent through the Earth's atmosphere and falls onto the Moon.

Tides

▲ *Because of its shape, the 250-kilometer (150-mile) long Bay of Fundy in Canada has a bigger difference between tides than anywhere else on Earth – high tide there is nearly 15 meters (50 feet) higher than low tide. These photographs were taken at high and low tide at Hopewell Cape, on the New Brunswick side of the bay.*

Tides are caused by the gravity of the Sun and Moon pulling on the oceans of the Earth, creating two bulges of water. The Moon's tidal pull is greater than the Sun's because it is much closer to us. The bulges in the oceans are highest when the Moon and Sun are pulling in line, at full Moon and new Moon. But when the Moon and Sun are at right angles to Earth, at first and last quarter, the bulges are not so high. The oceans rise and fall as the Earth rotates under the bulges of water, so that most places have two high tides and two low tides each day.

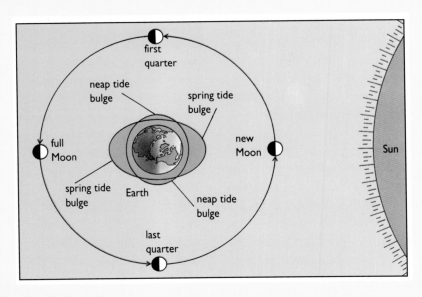

Mercury

Of all the planets, closest to the Sun is Mercury, an airless ball of rock less than half the size of the Earth. It is difficult to see because it never strays far from the Sun. Until the 1960s astronomers knew little about it. For example, they thought that the planet turned on its axis every 88 days, which is also the time it takes to orbit the Sun. This would mean that one side of it always faced the Sun, while the other side always faced away, so the Sun would never rise or set as seen from the surface.

But in 1965 there came a surprise. Astronomers bounced radio waves off Mercury and, from the echoes, they could tell how fast the planet spins. The result was 59 days, two-thirds of the time it takes to orbit the Sun. So the Sun does rise and set on Mercury, but very slowly. One "day" on Mercury – say, from one noon to the next – lasts 176 Earth days. In that time, Mercury orbits the Sun twice and spins three times on its axis.

Few markings can be seen on Mercury, even through large telescopes. The first good look at its surface came from the US space probe Mariner 10, which flew past Mercury in 1974. Mariner 10's cameras showed that Mercury looks a lot like our Moon. Mercury's surface is covered with craters up to 400 kilometers (250 miles) across. Like the craters on the Moon, they were formed when large meteorites smashed into the planet's surface long ago.

There may be surprises to come on the parts of Mercury not seen by Mariner 10. A US probe called Messenger, to be launched in 2004, and a European probe called BepiColombo, due for launch in 2009, will each go into orbit around Mercury, photographing its entire surface in detail and studying its composition.

In the daytime Mercury's surface is

▲ A mosaic of photographs from the space probe Mariner 10, whose path took it close to Mercury twice in 1974 and once again in 1975. On these three visits, each lasting just a few hours, Mariner 10's cameras took pictures of half the planet's surface. Future space probes will finish the job.

seared by the Sun's rays, heating up to over 400°C (750°F), enough to melt tin and lead. At night, however, the temperature plunges to –170°C (–275°F). Mercury is a hostile world and no one is likely to visit it for a long time, if ever.

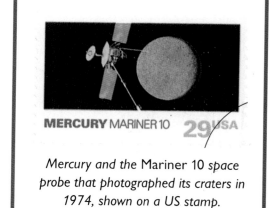

MERCURY MARINER 10   29 USA

Mercury and the Mariner 10 *space probe that photographed its craters in 1974, shown on a US stamp.*

MERCURY DATA

Distance from Sun:	36 million mi
Diameter:	3030 mi
Time to orbit Sun:	88 days
Time to spin on axis:	59 days
Mass:	0.06 × Earth
Volume:	0.06 × Earth
Average density:	5.4 × water
Tilt of axis:	0°
Number of moons:	0

Transits of Mercury

Sometimes the tiny disk of Mercury can be seen through a telescope crossing the face of the Sun. This event is called a *transit*. Transits do not happen often: the next transits of Mercury are in 2003, 2006, 2016 and 2019. This sequence of pictures of Mercury in transit was taken in November 1999 by a Sun-watching satellite called TRACE.

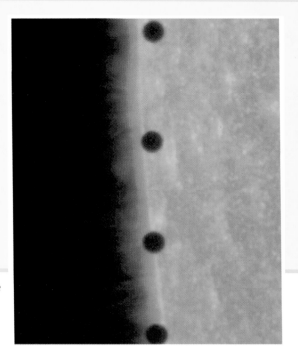

▼ Map of Mercury, drawn from photographs taken by Mariner 10, covering one hemisphere (but not the polar regions). Parts the probe's cameras could not see are left white. Most lowland areas (called planitiae) have as their name the word for Mercury in different languages. The craters are named after famous composers, artists, and writers.

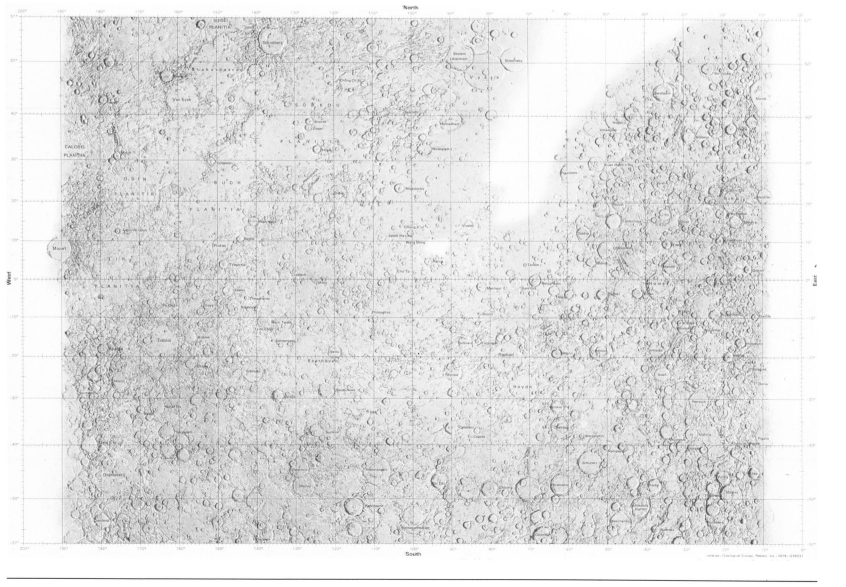

Venus, the Hell planet

Venus is easy to spot when it shines brilliantly in the twilight as the morning or evening "star" – far brighter than any true star. A small telescope shows that Venus is not a star at all, but a planet that goes through a cycle of phases as it orbits the Sun (see the diagram).

However, no telescope can show the surface of Venus because the planet is totally wrapped in clouds. It is sunlight reflected off these clouds that makes the planet so bright. Astronomers could only guess what lay under the clouds of Venus until space probes reached the planet. Although Venus is almost the same size as the Earth, in many other ways the two planets are very different.

The first space probe to fly past Venus, *Mariner 2* in 1962, found that the planet was scorchingly hot. Later probes landed on the planet's surface, and recorded oven-like temperatures of 475°C (890°F), even at night! The reason for these high temperatures is the atmosphere of Venus. It is almost entirely carbon dioxide – the gas we breathe out. Carbon dioxide traps heat from the Sun very well, so underneath the clouds of Venus it has got hotter and hotter. This is an extreme example of what is called a *greenhouse effect*. Scientists are worried that a similar greenhouse effect will heat up the Earth if we let too much carbon dioxide from burning fuels get into our atmosphere.

As well as being baking hot and unbreathable, the atmosphere of Venus bears down on the planet's surface with a crushing force of 90 times the Earth's atmospheric pressure. No wonder the Soviet *Venera* probes that parachuted down to its surface soon stopped working.

Adding to the Hellish nature of Venus is the composition of the clouds themselves. On Earth, clouds are made of water vapor, but there is no water on Venus. Its clouds consist of droplets of sulfuric acid, the same type of acid that is in a car battery. Who would ever want to go to Venus?

▲ Artist's impression of a Venera probe on Venus. These probes took photographs and studied the surface rocks. The strange orange light is produced by the clouds of sulfuric acid around Venus. Huge flashes of lightning may crackle through the murky air.

Venus occasionally crosses the face of the Sun as seen from Earth, an event known as a transit. Transits of Venus are rare, and were once regarded as so scientifically important that Captain James Cook was sent to the Pacific Ocean in 1769 to observe one, as commemorated on this stamp from Norfolk Island.

   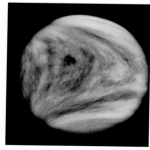

◀ The clouds of Venus spiral from the equator to the pole, as seen in this series of photographs in ultraviolet light from the Pioneer Venus Orbiter probe. The cloud tops lie 65 kilometers (40 miles) above the planet's surface.

► On its way to Mercury, Mariner 10 passed Venus and took this photograph in ultraviolet light. Some people looking at Venus through ordinary telescopes believe they have seen similar markings. (The blue color is not real, it has been added by a computer process to help show the detail in the clouds.)

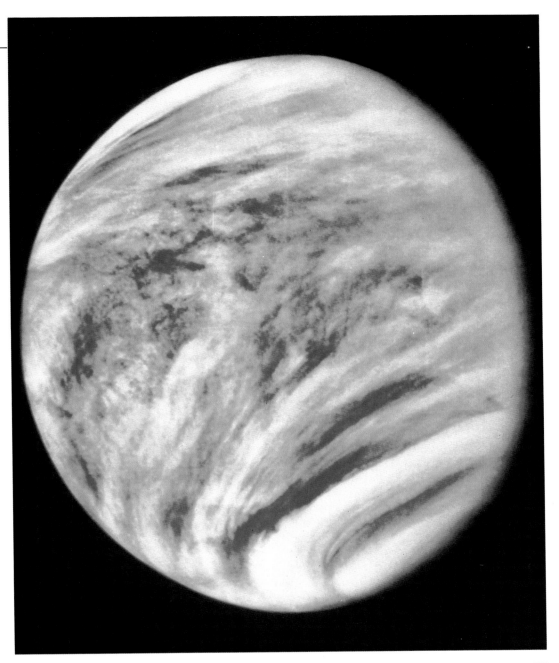

VENUS DATA

Distance from Sun:	67 million mi
Diameter:	7520 mi
Time to orbit Sun:	225 days
Time to spin on axis:	243 days
Mass:	0.82 × Earth
Volume:	0.86 × Earth
Average density:	5.24 × water
Tilt of axis:	177°
Number of moons:	0

▼ As Venus orbits the Sun, it goes through phases similar to those of the Moon. Venus is at its brightest when it is a crescent. At its closest, Venus lies 40 million kilometers (25 million miles) from the Earth, closer than any other planet — although then it is between us and the Sun and cannot be seen.

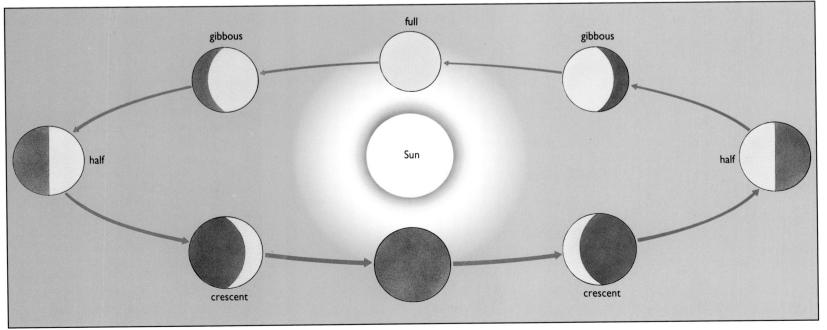

The surface of Venus

Beneath its clouds, Venus is a world with continents, rolling plains, canyons, volcanic mountains, and meteorite craters. These features have been revealed by using radar to "see" through the clouds, from space probes and from radio telescopes on Earth. The most detailed survey was made by the US space probe *Magellan* that went into orbit around Venus in 1990, although the main features were already known from previous radar mapping.

The highest point on Venus is a mountain range in the northern hemisphere, Maxwell Montes, named after a Scottish scientist. Its summit towers 11 kilometers (7 miles) above the average surface level of Venus. (Because there is no water on Venus there is no such thing as "sea level.")

Maxwell Montes is part of a continent on Venus called Ishtar, the size of Australia. The largest continent on Venus, half the size of Africa, is Aphrodite, on the planet's equator. It includes the second-highest peak on Venus, Maat Mons, 8.5 kilometers (5 miles) high. Maat Mons is believed to be an active volcano, and it is likely that other mountains on Venus were formed by volcanic eruptions. The Soviet *Venera* probes that landed on Venus found that its surface rocks were similar to volcanic lava on Earth.

As well as volcanoes there are craters where large meteorites hit the planet. Even Venus's dense atmosphere is no protection against the largest meteorites.

Nearly all the features on Venus are named after women, real or mythical. For example, Ishtar and Aphrodite (after whom the main continents are named) were both goddesses, but many of the craters are named after women scientists and artists.

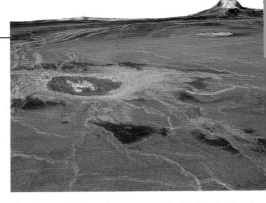

▲ *Part of Venus known as Eistla Regio. In the middle of the picture is a crater called Cunitz, 48 kilometers (30 miles) in diameter, and in the distance is Gula Mons, a volcano.*

▼ *Gula Mons from closer up. This volcano is 3 kilometers (2 miles) high, and it is surrounded by lava flows, hundreds of kilometers long, on plains covered with cracks. Cracks like this can be seen in the other pictures too.*

▼ *Maat Mons, an 8.5-kilometer (5-mile) high volcano on Venus's equator that might still be active. The pictures on these two pages were made by computer-processing parts of Magellan's radar maps to create three-dimensional "views" of the surface. The color has been added to make them look more real, and the height of mountains has been "stretched."*

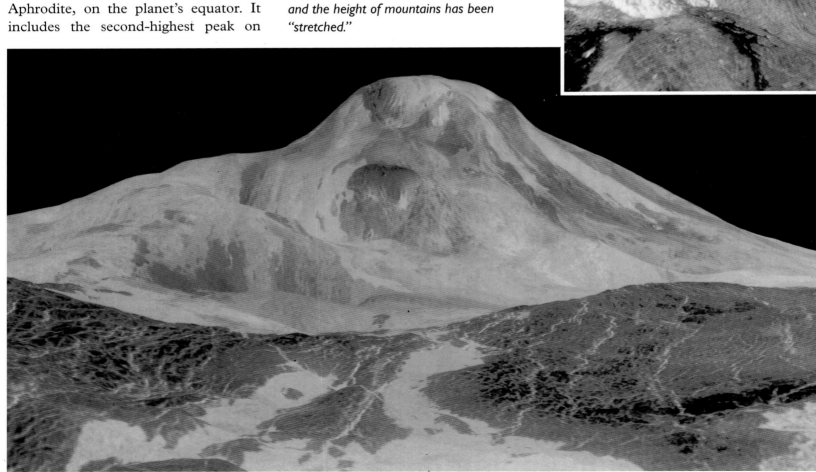

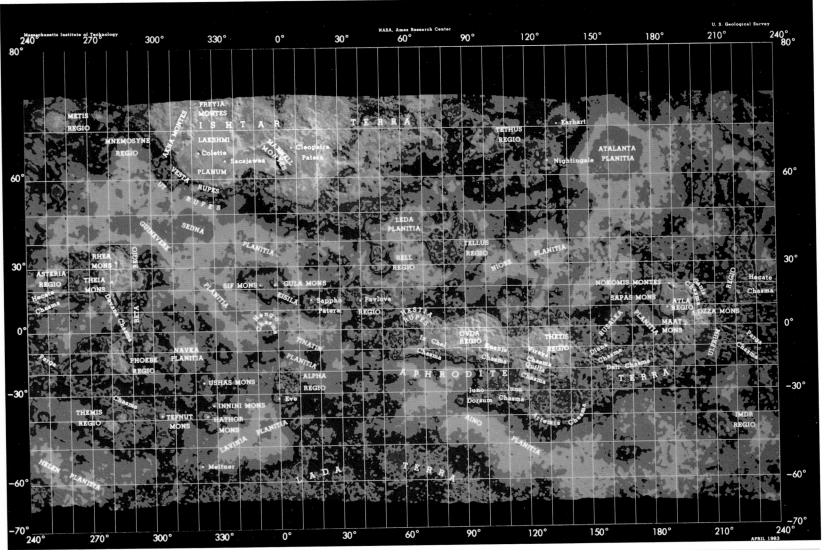

Map of Venus, showing surface features labeled across a grid with longitude marked at top and bottom (240°, 270°, 300°, 330°, 0°, 30°, 60°, 90°, 120°, 150°, 180°, 210°, 240°) and latitude marked on sides (80°, 60°, 30°, 0°, -30°, -60°, -70°).

Massachusetts Institute of Technology NASA, Ames Research Center U. S. Geological Survey

Labels on map include: METIS REGIO, MNEMOSYNE REGIO, ANNA MONTES, FREYJA MONTES, ISHTAR TERRA, LAKSHMI, Colette, Sacajawea, PLANUM, MAXWELL MONTES, Cleopatra Patera, VESTA RUPES, UT RUPES, RUPES, SEDNA, Earhart, TETHUS REGIO, ATALANTA PLANITIA, Nightingale, ASTERIA REGIO, RHEA MONS, THEIA MONS, Hecate Chasma, GUINEVERE, BETA REGIO, Devana Chasma, SIF MONS, GULA MONS, EISILA, Sappho Patera, Pavlova, REGIO, LEDA PLANITIA, BELL REGIO, TELLUS REGIO, NIOBE, PLANITIA, NOKOMIS MONTES, SAPAS MONS, ATLA REGIO, OZZA MONS, Hecate Chasma, Parge Chasma, Parge, PHOEBE REGIO, NAVKA PLANITIA, Nemo Chasma, TINATIN PLANITIA, HESTIA RUPES, Lo Chei Chasma, OVDA REGIO, Kuanja Chasma, THETIS REGIO, Viraye Chasma, Quilla Chasma, RUSALKA PLANITIA, MAAT MONS, Diana Chasma, Dali Chasma, Cani Chasma, Parga Chasma, ULFRUN REGIO, THEMIS REGIO, USHAS MONS, INNINI MONS, Eve, ALPHA REGIO, APHRODITE TERRA, Juno Dorsum, Juno Chasma, Artemis Chasma, TERRA, IMDR REGIO, TEFNUT MONS, HATHOR MONS, LAVINIA PLANITIA, AINO PLANITIA, Meitner, LADA TERRA, HELEN PLANITIA

APRIL 1983

▲ *Map of Venus, colored — like maps of the Earth — to show the height of the ground, from blue (lowest) up to red and gray (highest). As on Mercury, the lowland plains are called* planitiae. *Other words on the map are* mons *(mountain),* regio *(region), and* chasma *(canyon).*

▼ *Three craters on the surface of Venus. The one in front is Howe (diameter 37 kilometers/ 23 miles). The other two are called Danilova (48 kilometers/30 miles), on the left, and Aglaonice (64 kilometers/39 miles). Because of the extreme surface conditions on Venus, craters like these have been worn down and look "flat" compared with craters on the Moon.*

The strange spin of Venus

Venus spins very slowly on its axis, once every 243 days. This is even longer than the time it takes to orbit the Sun, 225 days. It is the only planet that takes longer to spin than to orbit the Sun. Another odd fact is that it spins from east to west, the opposite direction to the spin of the Earth and other planets. No one knows why the spin of Venus should be so strange. Perhaps another body hit it long ago, reversing its spin. The clouds of Venus also move around the planet from east to west, but they do so every four days, blown by howling winds of 360 kilometers (220 miles) per hour.

Mars, the red planet

Mars was once thought to be the most likely planet on which we would find other life. Although it is only half the size of Earth, in other ways it seems like our home planet. The day on Mars is only just over half an hour longer than ours, and it has an atmosphere with clouds. There are polar caps that melt in summer and spread in winter, and deserts too. Most interesting of all are markings that change slightly in size and shape each year, rather as if plants were growing. Some astronomers even thought they could see "canals" dug by Martian beings to bring water from the poles to the deserts.

Space probes have shown us many interesting things on Mars, but no life has yet been found there. For a start, it is too cold. Even on a summer's day, the temperature never rises above freezing point. What's more, the air is as thin as the Earth's air at a height of about 35 kilometers (22 miles), far too thin to breathe. Enough ultraviolet radiation from the Sun reaches the surface to kill anything living.

There is water on Mars, but it is not liquid. It is frozen away beneath the surface and in the polar caps. But Mars probably had liquid water in the past.

Some markings on the planet's surface look like dried-up river beds. Perhaps Mars was warmer in the past, so that the ice melted for a time.

The river beds on Mars are not the "canals." Those seem to have been a trick of the eye, an optical illusion, that fooled astronomers as they looked through their telescopes. The dark areas, once thought to be plants, are simply areas of darker rock and dust. They change in shape each year because the dust is blown around by the winds.

Exploring the surface of Mars

In July 1997 an American space probe, Mars Pathfinder, landed on Mars carrying a small roving vehicle called Sojourner (above). The pictures from Mars Pathfinder showed the rocky red surface of the planet. There is a lot of iron in the rocks, and it has combined with oxygen from the air to make rust. At the end of 2003, a European space probe called Mars Express will go into orbit around Mars and a small lander called Beagle 2 (left) will drop to the surface. Beagle will drill out a sample from under a rock and analyse it for signs of simple life forms. Future probes will bring samples of Martian rocks back to Earth for scientists to study. One day, humans will travel to Mars to explore its red deserts for themselves.

► Mars as seen from Earth, photographed through the Hubble Space Telescope. The red deserts and dark markings can clearly be seen. The long dark "tongue" to the right of center is a plain called Syrtis Major. This is one of the areas that was once thought to be covered with plant life. To the south of it, hazy white clouds cover a lowland area called Hellas. At the top of the picture is the planet's north polar cap, which has shrunk in the northern hemisphere summer.

◄ Mars, photographed from an approaching Viking space probe.

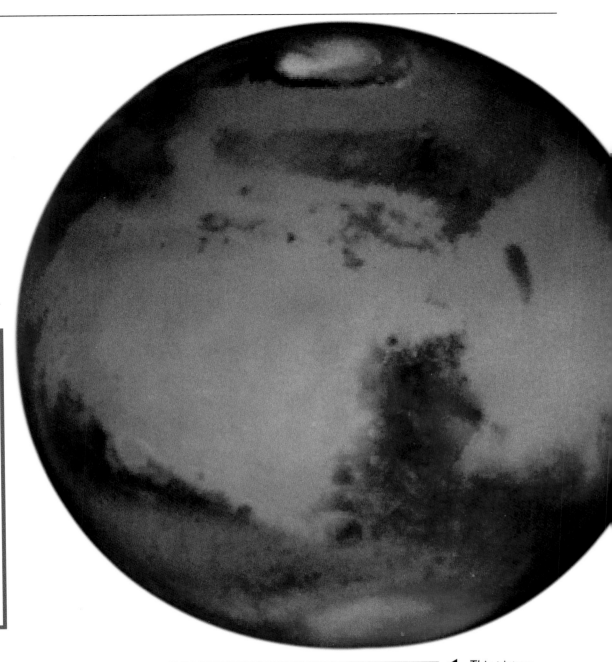

This American stamp shows one of the two Viking space probes that reached Mars in 1976. Each probe came in two halves, one of which landed on the planet, while the other remained in orbit.

MARS DATA

Distance from Sun:	142 million mi
Diameter:	4220 mi
Time to orbit Sun:	687 days
Time to spin on axis:	24 hr 37 min
Mass:	0.11 × Earth
Volume:	0.15 × Earth
Average density:	3.94 × water
Tilt of axis:	25.2°
Number of moons:	2

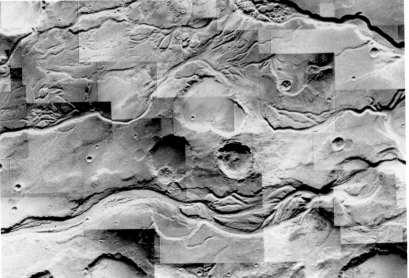

◄ This picture is made up of photographs taken by one of the Viking orbiters. It shows what might long ago have been rivers flowing with water. Other Viking photographs show what astronomers think are ancient sea beds and areas where there were floods.

The volcanoes and canyons of Mars

Tourists who visit Mars next century will want to see two things: its huge volcano, Olympus Mons, and a vast canyon system called the Mariner Valleys. Olympus Mons is larger than any volcano on Earth. It is shaped like a dome, 600 kilometers (375 miles) wide at its base and about 25 kilometers (15 miles) high, which would dwarf Mount Everest. At its top is a crater 90 kilometers (55 miles) wide, big enough to swallow several cities. Near Olympus Mons is a row of three smaller (but still very large) volcanoes in an area called the Tharsis Ridge.

To the east of the Tharsis Ridge are the Mariner Valleys (or Valles Marineris), named after the Mariner 9 spacecraft, the first to photograph them. The whole group is over 4000 kilometers (2500 miles) long, enough to stretch across the United States. In the middle it is 7 kilometers (4.5 miles) deep and 600 kilometers (375 miles) wide.

Although it is called a "canyon," it is not like the Grand Canyon in the United States, which was carved out by a river and is much smaller. Instead, Valles Marineris is more like the Rift Valley of East Africa. It was formed by cracking in the crust of Mars. Winds and running water in the past have helped to make the canyons of Mars larger.

Other parts of Mars look like the Moon. Most of the planet's southern hemisphere, for example, is highlands with craters made by meteorites, not volcanoes. One large feature in the southern hemisphere is like a lunar "sea." Called Hellas, it is a lowland plain 1800 kilometers (1100 miles) wide and 3 kilometers (2 miles) deep. It is thought to have been dug out by a meteorite that hit Mars 3500 million years ago. Other large impact basins are Argyre, also in the southern hemisphere, and Isidis, near the Martian equator.

◄ *This mosaic of photographs taken from the Viking space probes shows nearly half the surface of Mars. Running from left to right is the great system of valleys called Valles Marineris. What is probably the dried-up bed of an ancient river winds its way northward from the valleys. On the left is the Tharsis Ridge, with its three giant volcanoes.*

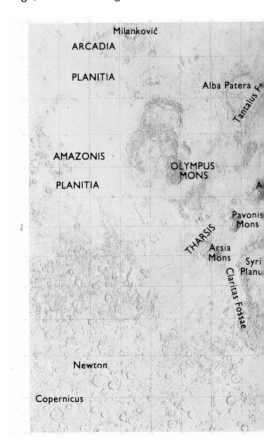

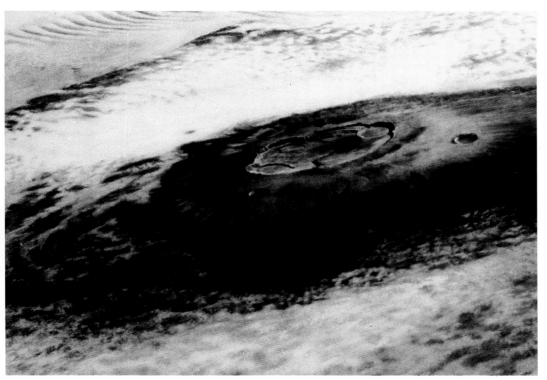

The moons of Mars

Mars has two moons, Phobos (pictured left) and Deimos. They are both small and oddly shaped, like lumpy potatoes. Phobos is only 28 kilometers (17 miles) at its widest, and Deimos is 16 kilometers (10 miles). Both moons have been cratered by meteorites. Phobos and Deimos are thought to be ex-asteroids that were captured into orbit around Mars.

▼ Map of Mars, showing some of the main features. Lowland plains are called planitiae. Many consist of lava flows, like the maria on the Moon. Volcanoes have the word mons or patera in their name. A planum is an upland plain. Fossae are long grooves.

▲ The summit of Olympus Mons, surrounded by cloud. Olympus Mons is the biggest volcano so far discovered in the Solar System. The mountain and its clouds can be seen through a telescope from Earth. It used to be called Nix Olympica, "The snow of Olympus."

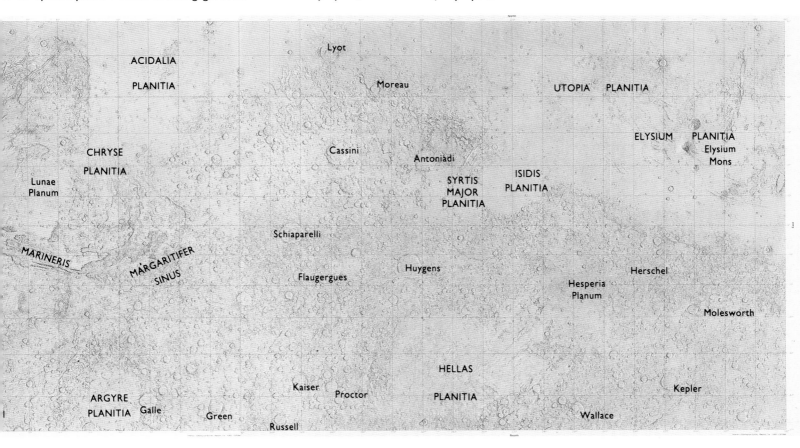

Jupiter, the giant planet

◀ *Jupiter as photographed from Earth with the Hubble Space Telescope, showing its colorful cloud bands. The moon Io is visible just above center, with its round black shadow to the left.*

JUPITER DATA

Distance from Sun:	484 million mi
Diameter:	88,800 mi
	(at the equator)
Time to orbit Sun:	11.9 years
Time to spin on axis:	9 hr 50 min
	(at the equator)
Mass:	318 × Earth
Volume:	1323 × Earth
Average density:	1.3 × water
Tilt of axis:	3.1°
Number of moons:	28

Beyond Mars and the asteroid belt, the planets change from being small and rocky to large and made of gas. Largest of all is Jupiter, eleven times wider than the Earth and over twice the weight of all the other planets put together. It never comes closer than about 600 million kilometers (370 million miles) from Earth, but because of its large size it still shines more brightly than any star in the night sky.

Jupiter is mostly made of hydrogen and helium. On Earth, hydrogen and helium are gases. But under the tremendous pressures inside a planet as big as Jupiter they are squeezed into a liquid. So Jupiter is mostly a ball of liquid hydrogen and helium, covered with a layer of clouds. There is no solid surface for a spaceship to land on.

Jupiter's clouds are drawn out into bands by the planet's fast spin – its "day" lasts less than 10 hours. Jupiter bulges at the equator because it spins so quickly. Various chemicals turn the clouds white, yellow, brown, and red in color.

The weather on Jupiter is stormy, and the clouds keep changing in appearance as they swirl around the planet. Only one feature, a big storm cloud called the Great Red Spot, seems to be there all the time. In 1979 two American space probes called *Voyager* flew past the planet and photographed the clouds in detail. They also saw brilliant flashes of lightning on the night side of Jupiter, caused by thunderstorms.

In 1995 a space probe called *Galileo* reached Jupiter and went into orbit. It released a smaller probe that parachuted down through the clouds, telling us more about the winds and temperatures there. The probe was eventually crushed when the pressure became too great for it to withstand.

Jupiter's Great Red Spot

In Jupiter's southern hemisphere is a large cloud shaped like an eye. It is called the Great Red Spot, and was first seen when astronomers turned their telescopes on the planet in the 17th century. The Great Red Spot is about 25,000 kilometers (15,000 miles) long and 12,000 kilometers (7500 miles) wide, although it changes slightly in size and shape with time. It is big enough to swallow several Earths. The spot's red color is thought to come from phosphorus, the same substance as in the heads of red matches. It seems to be a spinning storm cloud, caused by gas rising from Jupiter's warm interior.

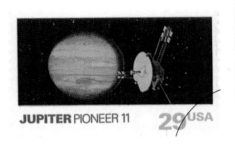

Our best views of Jupiter have come from the American space probe called Galileo, *which went into orbit around the planet in 1995. An earlier US probe called* Pioneer 11 *passed Jupiter in 1974 and is shown on this stamp.*

▶ *An artist's impression of the* Galileo *probe parachuting down into Jupiter's atmosphere.*

◀ *In this* Voyager *photograph of Jupiter's clouds, the colors have been exaggerated to show more detail. Winds in the atmosphere blowing at up to 450 kilometers (275 miles) per hour create these complicated and beautiful patterns.*

The moons of Jupiter

Jupiter has 28 known moons, and there may be others too small to have been seen yet. The four biggest and brightest moons were discovered by the Italian scientist Galileo Galilei in 1610. An ordinary pair of binoculars shows them as tiny points of light, moving around the planet from night to night.

In order of distance from the planet, they are called Io, Europa, Ganymede, and Callisto. Ganymede and Callisto are bigger than our own Moon and the other two are not much smaller. Of the four, Io is the most interesting because it has active volcanoes which have been photographed erupting by the Voyager and Galileo space probes. Plumes of sulfur and sulfur dioxide are sprayed upwards. Sulfur from the volcanoes gives the surface of Io a peculiar orange color. Io is thought to be kept hot by Jupiter's strong gravity, which "squeezes" the moon, releasing heat.

By contrast, the next moon, Europa, is covered with white ice. Most of its surface is smooth, but in places it is cracked like an eggshell. Ganymede is the largest moon in the Solar System, and is bigger than the planet Mercury. Callisto is the same size as Mercury. Both Ganymede and Callisto are peppered with craters caused by meteorites. As well as craters, Ganymede has strange grooves on its surface, probably due to movements in the crust. The surfaces of these moons are not rocky, like our own Moon, but consist of dirty ice.

All the other moons of Jupiter are much smaller than the big four. The smallest of Jupiter's moons are only about 3 kilometers (2 miles) across.

▶ *Every part of Callisto is covered with craters of all sizes. Some are very bright, showing that Callisto has an icy surface. The biggest impact on this moon was in an area called Valhalla, to the right in the close-up.*

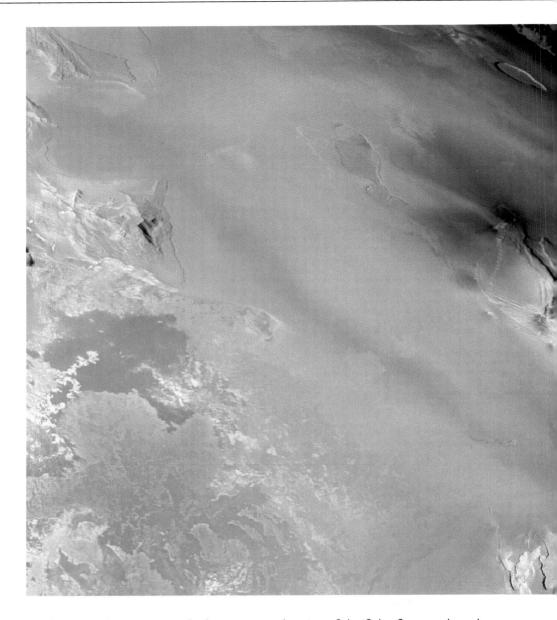

▲ *Of all the discoveries made during our exploration of the Solar System, the volcanoes of Io were the biggest surprise. This one, named Pele after a mythical god of fire, is spraying a great plume of gas and dust to a height of 300 kilometers (nearly 200 miles). The photograph was taken from the Voyager 1 space probe, in March 1979. Repeated eruptions of Pele have been photographed by the Galileo space probe.*

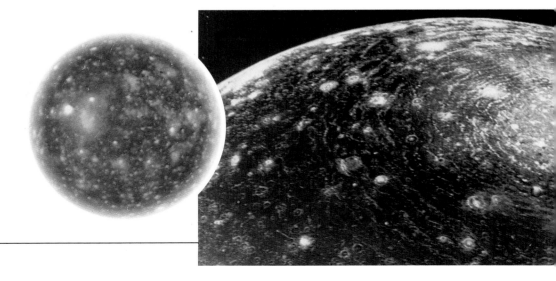

JUPITER'S MAIN MOONS DATA

Io
Diameter:	2250 mi
Distance from Jupiter:	262,000 mi
Time to orbit Jupiter:	1.8 days

Ganymede
Diameter:	3270 mi
Distance from Jupiter:	660,000 mi
Time to orbit Jupiter:	7.2 days

Europa
Diameter:	1950 mi
Distance from Jupiter:	417,000 mi
Time to orbit Jupiter:	3.6 days

Callisto
Diameter:	2980 mi
Distance from Jupiter:	1.2 million mi
Time to orbit Jupiter:	16.7 days

◀ *Europa has a very bright, icy surface. Seen in close-up, below left, it is criss-crossed by a network of lines, some stretching for hundreds of kilometers. These may be places where, long ago, water flowed up from inside to fill cracks in the surface, then froze.*

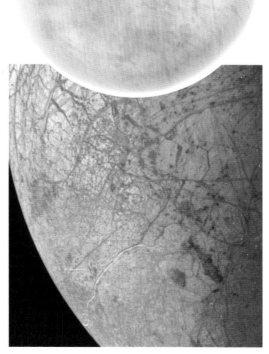

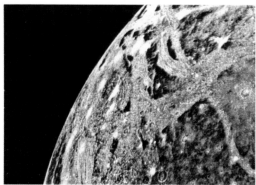

◀ *Jupiter's equator is circled by a faint ring of dust, discovered by the Voyager space probes in 1979. It is far too faint to be seen from Earth. The ring lies about 50,000 kilometers (30,000 miles) above the planet's cloud tops.*

▲ *Ganymede also has an icy surface, but with some dark areas and bright craters. The largest dark area is named after Galileo. Strange ridges and grooves are also found on this moon, as the close-up view shows.*

Saturn, the ringed planet

Saturn and its rings are one of the most beautiful sights in the night sky when seen through a telescope. Saturn is like a smaller version of Jupiter, although it does not have the same colorful bands of cloud in its atmosphere. There are cloud features, but they seem to be masked by a high-altitude layer of haze that gives the whole planet a smooth, yellowish look.

Every 30 years or so, a large white spot appears on Saturn. The last one, in 1990, was followed by several smaller outbreaks. These spots are actually bright clouds that form during summer in the planet's northern hemisphere.

Summer comes only every 30 years on Saturn because that is how long the planet takes to orbit the Sun.

Thirty moons have been discovered around Saturn, more than for any other planet, and there are probably other small ones yet unseen. The largest of them, Titan, is the only moon in the Solar System with an atmosphere to speak of. The other moons are mostly chunks of ice and rock. One, called Mimas, has a huge crater on it. The meteorite that caused the crater must almost have broken Mimas apart. Another odd-looking moon, Iapetus, has one bright side, covered with ice, and one dark side, coated with dust or perhaps other material. No one knows for certain why its two halves are so different.

SATURN DATA

Distance from Sun:	887 million mi
Diameter:	74,900 mi
	(at the equator)
Time to orbit Sun:	29.5 years
Time to spin on axis:	10 hr 14 min
	(at the equator)
Mass:	95 × Earth
Volume:	750 × Earth
Average density:	0.7 × water
Tilt of axis:	27°
Number of moons:	30

▲ Saturn and its magnificent rings. The Earth would fit three times over into the gap between the inner ring and the planet's surface. Part of Saturn's shadow falls on the rings in this photograph.

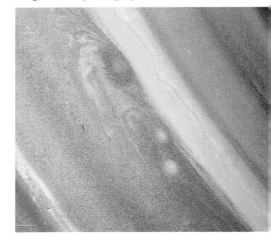

Saturn's moon Mimas, 390 kilometers (240 miles) wide, has a huge crater called Herschel, one-third its own width. The crater has a big central peak.

▼ One side of the moon Iapetus is bright and cratered, but the other side is dark. It may be sweeping up dust particles as it travels around its orbit. Iapetus is 1440 kilometers (890 miles) in diameter.

◀ Saturn's cloudy atmosphere is not as dramatic as Jupiter's, but it does show the same pattern of bands. Color has been added to this photograph to bring out the detail.

Titan

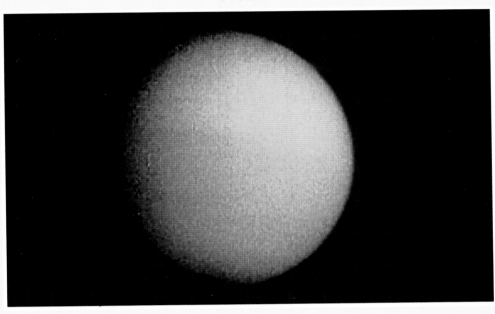

Titan, Saturn's main moon, is the second-largest moon in the Solar System. Only Jupiter's biggest moon, Ganymede, is larger. But what makes Titan unique among moons is that it is the only one to have a thick atmosphere. This atmosphere is mostly nitrogen, the main gas in our atmosphere on Earth, plus some methane. Titan has been called a "deep-freeze" version of the Earth. We cannot see the surface of Titan because of smoggy clouds, which give it an orange appearance, as seen in this Voyager 1 image (above). But Titan is so cold, because it is so far from the Sun, that methane may fall as rain on the surface. A space probe called Huygens is due to land on Titan in November 2004. This probe is part of the joint US–European space mission to Saturn named Cassini after an astronomer who studied the planet in the 17th century.

TITAN DATA	
Diameter:	3200 mi
Distance from Saturn:	760,000 mi
Time to orbit Saturn:	16 days

▲ Dione, 1120 kilometers (700 miles) wide, has a mixture of lightly and heavily cratered areas, surface cracks, and wispy marks that may be frost.

▼ The most prominent features in Saturn's clouds are white spots that break out from time to time near the planet's equator. This one was photographed by the Hubble Space Telescope in 1994.

Saturn's rings

Around Saturn's middle lies a set of bright rings. At first sight the rings look solid. But in fact they are made up of countless millions of frozen lumps that range in size from icebergs to snowballs. They all orbit Saturn like a swarm of tiny moonlets.

The rings measure 270,000 kilometers (170,000 miles) from side to side, over twice the width of the planet itself. Yet they are no more than a few hundred meters thick. In relation to their width, they are really as thin as a sheet of paper the size of a football field.

As seen from Earth, the rings seem to divide into three main bands, each of different brightness. The brightest part is in the middle, called the B ring. Outside it is the A ring, and between them is a gap called Cassini's Division, the width of the Atlantic Ocean. Closest to the planet is the faintest ring of all, called the C ring or crepe ring.

A completely new look at the rings was given by the *Voyager* space probes

▼ You can see the surface of Saturn through its rings in the photograph, which proves how thin they are.

Cassini is a joint US–European space probe to Saturn, launched in 1997. One part of the probe will go into orbit around Saturn in 2004. Another part, called Huygens, will land on Saturn's largest moon, Titan.

that flew past the planet. In close-up, the smooth-looking rings broke up into thousands of thin ringlets, like threads. In photographs like the one on the right, Saturn's system of rings looks like an enormous gramophone record. Even the Cassini Division contains some ringlets. Some other faint rings that cannot be seen from Earth were discovered inside the C ring and beyond the A ring.

Most fascinating of these is the F ring, which looks twisted. This is thought to be due to the effects of small moons that orbit either side of it. The

▶ The strange F ring (left), twisted, split, and uneven. This is caused by two tiny moons, called shepherd moons, which orbit either side of the F ring and keep it from straying. One of them, called Prometheus, is shown on the right, photographed against Saturn's disk. The F ring is the faint line below it.

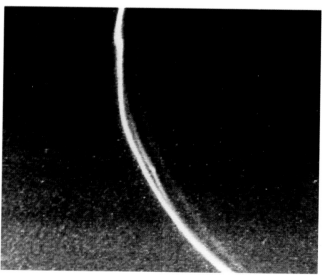

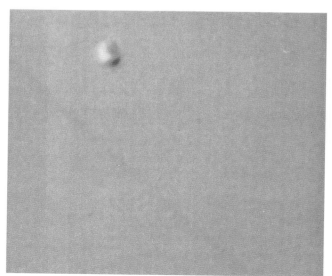

gravity of these small moons pulls the ring particles out of their normal orbits, giving the ring its odd shape.

How were the rings of Saturn formed? Even now, scientists are not sure. They could be material left over from the birth of the planet itself. Or they could be the remains of a moon that strayed too close to the planet and broke up. Other theories are that the ring particles came from collisions between moons, or are the remains of comets that crashed into moons, or broke up after being captured. Perhaps more than one of these causes gave rise to the rings we see today.

▲ *By a special computer process, color has been added to this* Voyager *photograph of Saturn's rings. The different colors may be telling us that the moonlets that make up the rings are made of slightly different material. The inner ring, the C ring, shows up blue, the B ring changes from orange to blue–green, and the outer A ring appears blue–gray.*

Spokes

Photographs taken by the *Voyager* space probes showed strange dark patches that come and go on Saturn's rings. Scientists named them "spokes." They are thought to be clouds of dust, possibly caused by the impact of meteorites on the rings.

Uranus, the tilted planet

Uranus was the first planet to be discovered with a telescope. At its brightest it can just be seen with the naked eye, like a faint star, if you know where to look. But no one had ever noticed it until 1781 when William Herschel first spotted it through his telescope from his back garden in Bath, England. Uranus lies twice as far from the Sun as Saturn, so the discovery doubled the size of the known Solar System overnight.

Uranus is four times wider than the Earth and is covered in greenish clouds. The green color is caused by methane gas in its atmosphere. However, most of the planet's atmosphere is hydrogen and helium, the same as on Jupiter and Saturn.

There are very few markings in the clouds of Uranus. Even the Voyager 2 space probe, which flew past the planet in 1986, found few cloud features of note. Underneath the clouds there is thought to be a deep ocean of water, methane and ammonia.

One remarkable fact about Uranus is that it is tilted on its side. As a result, the Sun can at times appear overhead at the poles! Uranus probably got its tilt long ago, when it was hit by another body. The collision literally knocked Uranus over so that its axis of rotation lies almost in the plane of its orbit.

Uranus has 21 known moons, many of which were discovered by the Voyager 2 space probe. The most interesting moon is Miranda, which looks as though it has been broken apart in the past and the parts have gathered themselves together again. Uranus also has a number of narrow, dark rings around its equator.

▲ *The four largest moons of Uranus. From the left, and in order of distance from the planet, they are: Ariel (diameter 1160 kilometers/720 miles), Umbriel (1170 kilometers/715 miles), Titania (1580 kilometers/980 miles), and Oberon (1520 kilometers/940 miles). All are icy, cratered worlds. Ariel has many scarps and valleys. On Titania, there are valleys and fractures that cut right through craters. The dark surface of Umbriel has one bright spot, a 110-kilometer (70-mile) crater called Wunda.*

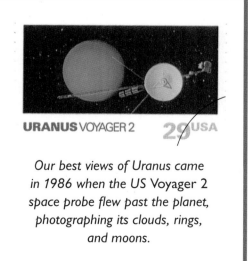

URANUS VOYAGER 2   29 USA

Our best views of Uranus came in 1986 when the US Voyager 2 space probe flew past the planet, photographing its clouds, rings, and moons.

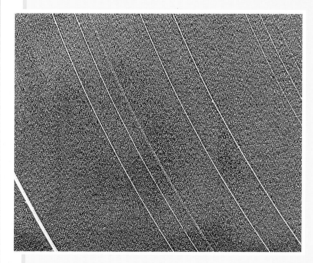

The rings of Uranus

Uranus is circled by a number of thin rings, too faint to see through telescopes on Earth except at infrared wavelengths. They were first discovered in 1977, when Uranus moved in front of a star. The rings made the star's light flash on and off as they passed across it. Eleven rings are now known, including two discovered by Voyager 2. Unlike the rings of Saturn they are as black as coal, which is why they are so difficult to see from Earth.

URANUS DATA

Distance from Sun:	1.78 billion mi
Diameter:	31,500 mi (equatorial)
Time to orbit Sun:	84 years
Time to spin on axis:	17 hr 14 min
Mass:	14.5 × Earth
Volume:	64 × Earth
Average density:	1.3 × water
Tilt of axis:	98°
Number of moons:	21

▶ Remarkable Miranda – no other body has such a dramatic mixture of surface types. There are grooves, ridges, and craters, as well as features not seen anywhere else, like the tick-shaped marking near the center, nicknamed "the chevron." Right at the bottom of the main picture, and shown in the close-up view, is a cliff that towers up to 20 kilometers (12 miles) high. Miranda is just 470 kilometers (290 miles) in diameter.

▲ Shrouded in greenish blue clouds – Uranus, photographed by Voyager 2. Unlike Jupiter and Saturn, Uranus's clouds show little sign of activity.

Neptune, the blue giant

Early last century, astronomers noticed that Uranus was not keeping to its expected orbit. They guessed the reason was that the gravity of an unknown planet was pulling it off course. Two men, John Couch Adams in England and Urbain Le Verrier in France, set out to calculate where that unknown planet might lie. Both men came up with similar answers. When a German astronomer, Johann Galle, looked near the predicted position in 1846, he found Neptune.

Neptune is four times the width of Earth, almost the same size as Uranus. Through a telescope Neptune appears blue–green. This color is caused by methane in its atmosphere, as is the color of Uranus.

Little was known about Neptune until the Voyager 2 space probe passed it in 1989. Neptune is so far away that Voyager 2's radio messages, traveling at the speed of light, took over four hours to reach Earth. The most striking feature among Neptune's blue–green clouds was a dark spot the size of the Earth. This spot is thought to be a storm cloud, similar in nature to the Great Red Spot of Jupiter.

Like Uranus, Neptune turned out to have faint rings. Voyager discovered six new moons around Neptune to add to the two already known, Triton and Nereid. Long ago, both Triton and Nereid may have orbited the Sun separately but were captured by Neptune's gravity when they strayed too close. Many small moons around other planets were probably captured like this.

▶ Neptune's blue disk. Neptune's atmosphere is more active than Uranus's. In this Voyager photograph are dark spots and white clouds of methane ice crystals that look like the cirrus clouds of Earth.

◀ Neptune's second-biggest moon, called Proteus, was discovered by Voyager 2. It is about 400 kilometers (250 miles) across, 50 kilometers or so bigger than Nereid. It takes just over a day to orbit Neptune.

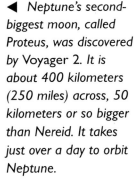

▶ Like the other giant planets, Neptune has rings round it. The two main ones are shown here. The photograph shows bright clumps in the outer ring.

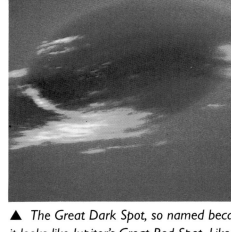

▲ The Great Dark Spot, so named because it looks like Jupiter's Great Red Spot. Like Jupiter's spot, it is probably a spinning storm cloud, with gas rising from below. It is about 12,000 kilometers (7500 miles) long.

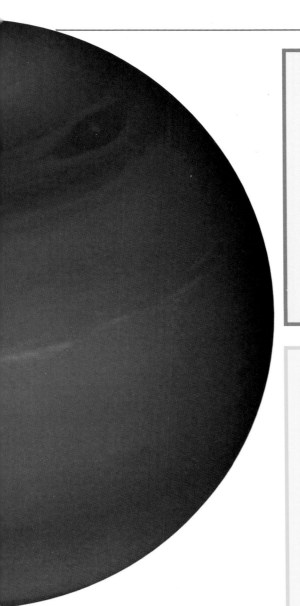

NEPTUNE DATA

Distance from Sun:	2.8 billion mi
Diameter:	30,200 mi (equatorial)
Time to orbit Sun:	165 years
Time to spin on axis:	16 hr 7 min
Mass:	17.2 × Earth
Volume:	54 × Earth
Average density:	1.8 × water
Tilt of axis:	28°
Number of moons:	8

NEPTUNE VOYAGER 2 **29**USA

Voyager 2 *reached Neptune in 1989 after a 12-year journey from Earth. During the journey it also passed three other planets: Jupiter, Saturn, and Uranus.*

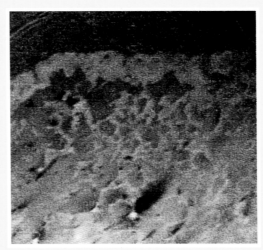

Triton

Neptune's largest moon, Triton, is covered with pink and bluish ice, as shown in these images by *Voyager 2*. The ice consists of frozen nitrogen and methane at a temperature of −236°C (−393°F), making it the coldest known place in the Solar System. Some of this ice evaporates to form a very thin atmosphere. Most astounding of all are the volcanoes of liquid nitrogen that erupt from pools beneath Triton's surface, spraying material nearly as high as Mount Everest. Where this material falls back onto Triton it produces dark streaks, as can be seen in the smaller picture.

TRITON DATA

Diameter:	1680 mi
Distance from Neptune:	220,000 mi
Time to orbit Neptune:	5.9 days

Pluto, at the edge of darkness

Pluto is the most recent planet to be discovered. It was found in 1930 by an American astronomer, Clyde Tombaugh, during a careful search for new planets beyond Neptune. But Pluto turned out to be very odd. Astronomers had expected to find another planet like Uranus or Neptune. But Pluto was much fainter, and therefore smaller, than expected. Not until recently did astronomers find out just how small Pluto really is. It turns out to be the smallest planet in the Solar System, smaller even than our own Moon and Neptune's moon Triton.

Another oddity about Pluto is its orbit. At times, Pluto can come closer to the Sun than Neptune, as it did between 1979 and 1999. Fortunately, there is no chance of a collision between Pluto and Neptune. Their orbits are like two hoops, one larger one tilted outside the other, so that they do not actually meet at any point.

We do not have any close-up photographs of Pluto because no space probe has been there and none is on its way. However, astronomers think that Pluto has ice on its surface and looks very much like Triton. In fact, both Pluto and Triton may once have orbited the Sun at the edge of the Solar System. Triton could have been captured by Neptune's gravity to become a moon, but Pluto remained free.

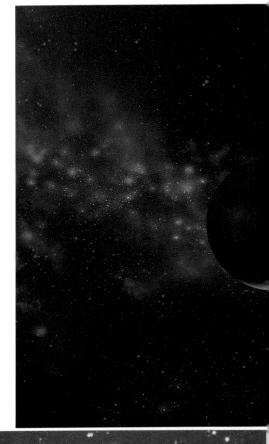

▼ *Clyde Tombaugh searched for a new planet by photographing a small part of the night sky, waiting a few days, and taking another photograph. A planet would show up by being in different places on the two pictures. Eventually he found what he was looking for on these two photographs of part of the constellation Gemini. The arrows point to Pluto. Tombaugh carried on looking for several years, but found no more planets.*

▼ We can only guess what Pluto and its moon look like until a space probe gets there. A Pluto probe might be launched to arrive by 2015. This artist's impression shows Charon in Pluto's sky, with the Sun just a bright star in the distance.

Other planets?

From time to time, some astronomers have speculated that unknown planets might lie beyond Pluto. However, any large planets like Uranus or Neptune would have been seen long ago. In recent years, astronomers have discovered a swarm of small, icy bodies 100 km (62 miles) or so across orbiting further from the Sun than Neptune. These are known as trans-Neptunian objects, and it seems that Pluto is simply the largest of them. It is unlikely that there are any undiscovered planets at the edge of the Solar System.

PLUTO NOT YET EXPLORED **29** USA

Pluto is the only planet that has not been reached by space probe.

Pluto's moon

Pluto has one moon, Charon, discovered in 1978. Charon is half the size of Pluto. This is larger in relation to the planet itself than any other moon (our own Moon, for example, is one-quarter the width of the Earth). Charon orbits Pluto every 6.4 days, the same time the planet takes to spin on its own axis. Therefore, the moon remains over one side of Pluto all the time and can never be seen from the other side of Pluto.

PLUTO DATA

Distance from Sun:	2.8 billion mi to 4.5 billion mi
Diameter:	1430 mi
Time to orbit Sun:	248 years
Time to spin on axis:	6.4 days
Mass:	0.002 × Earth
Volume:	0.01 × Earth
Average density:	1.1 × water
Tilt of axis:	123°
Number of moons:	1

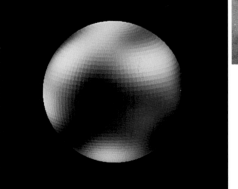

▲ *Pluto, seen through the Hubble Space Telescope. The bright and dark areas are thought to be caused by patches of frost on its surface. Close-up, Pluto probably looks like Neptune's largest moon, Triton.*

Comets, ghostly wanderers

At the edge of the Solar System, far beyond Neptune and Pluto, orbits a swarm of comets. They are frozen bodies just a few miles across, like dirty snowballs. There are uncountable numbers of them, but at that great distance they are invisible from Earth.

Occasionally, the gravity of a passing star nudges some of the dirty snowballs towards the Sun. As they approach the Sun they start to warm up. The ice turns into gas, and dust is released. The gas and dust forms a large cloud around the dirty snowball. This cloud is called the *coma*, and it can grow to become ten times the size of the Earth.

▼ *The three main parts of a comet: the coma, the nucleus, and the tail. Together, the nucleus and coma are known as the head.*

tail

nucleus

coma

At the center of the coma lies the dirty snowball, known as the comet's *nucleus*.

Once a comet develops a coma it becomes big and bright enough to be seen in telescopes from Earth. In some cases the gas and dust streams away to form a *tail*. A comet's tail can stretch for 10 to 100 times the distance from the Earth to the Moon. The tail is blown by the effects of sunlight and the solar wind so that it always points away from the Sun, no matter in which direction the comet is moving. As a comet moves between the planets, its path can be altered by their gravitational pull so that it comes back to the Sun regularly.

Each year about two dozen comets are seen. Some of these are new discoveries, while others are known comets returning to the Sun. The comet with the shortest orbit is Encke's Comet, which goes around the Sun every 3.3 years. But other comets take over a million years to go around the Sun. Most comets are named after the people who discover them. Many comets are discovered by amateur astronomers who spend their nights looking for them.

Every few years a comet becomes bright enough to be seen with the naked eye and puts on a ghostly show in the night sky. In the past, people thought that comets were bad omens and they feared them. Although a comet looks impressive, particularly in photographs, it is mostly composed of gas far thinner than the Earth's atmosphere. A comet would be dangerous only if its nucleus actually hit us.

► *Many photographs from the probe Giotto were combined to make this picture of the nucleus of Halley's Comet. Bright jets of dust and gas being squirted out go to form the comet's coma and tail.*

This stamp, issued by Great Britain, shows Edmond Halley looking something like his comet.

► Halley's Comet photographed from Chile in March 1986. People in the southern hemisphere had the best view of it. Although it was clearly visible, it was not as spectacular as at its last visit, in 1910. It is not due back again until the year 2062.

◄ Comet Hale–Bopp, a bright comet seen in 1997. The broad, white tail is made of dust, and the narrower blue tail is gas.

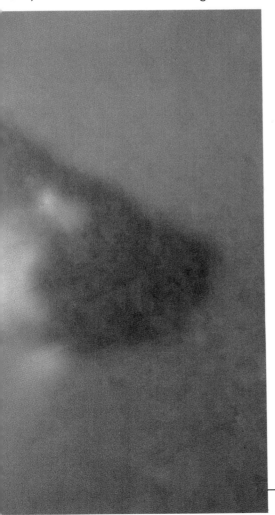

Halley's Comet

The most famous comet in history is named after an English astronomer, Edmond Halley, the first to realize that comets orbit the Sun. He suggested that the comets seen in 1531, 1607, and 1682 were all the same one, orbiting every 76 years or so. And he predicted that it would return again around 1758. True enough, the comet did come back as predicted and it was named after him.

Halley's Comet last appeared in 1986, when several spacecraft were sent to meet it. The European space probe *Giotto* went right through the comet's head, passing 600 kilometers (375 miles) from the nucleus. It took photographs showing the comet's nucleus, which looks like a lumpy potato. The nucleus is about 16 kilometers (10 miles) long and 8 kilometers (5 miles) wide, and is made of ice with a dark, dusty crust. This was the first time that a comet's nucleus had been seen.

▼ At its closest to the Sun, Halley's Comet comes inside the orbit of Venus. Its very elongated path takes it out beyond Neptune's orbit. (The innermost planets are not in this diagram.)

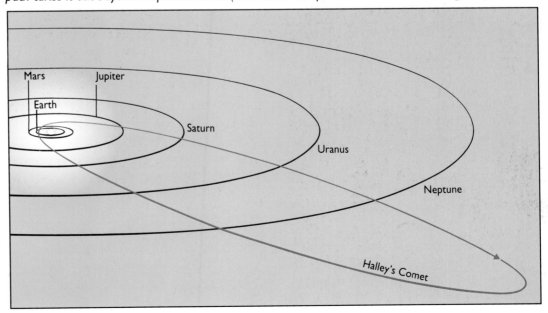

Asteroids, the cosmic killers

Asteroids, also called minor planets, are rubble left over from the formation of the Solar System. The first asteroid was found by chance in 1801 and named Ceres. Now over 10,000 are known, ranging from blocks the size of a mountain up to Ceres, the largest, which is over a quarter the size of our Moon. They are too faint to see without binoculars or a telescope. The brightest of them, Vesta, can sometimes be seen with the naked eye.

Most asteroids orbit in a region called the asteroid belt between Mars and Jupiter, but some stray through the inner Solar System. Learning more about such asteroids is important because, if one were found to be heading our way, we would need to divert or destroy it. With this in mind, in February 2000 a US space probe called NEAR Shoemaker went into orbit around the asteroid Eros, studying its composition and sending back images. After a year in orbit, it descended to the surface of Eros to make its final measurements.

▲ *In 1993 the American space probe Galileo took this photograph of the asteroid Ida, which it passed on its way to Jupiter. Ida turns out to be a lump of rock about 55 km (34 miles) long, pitted with craters. It has a tiny moon, Dactyl, seen on the right of this picture. Ida looks very much like the two moons of Mars, Phobos and Deimos, which are thought to be captured asteroids.*

Did an asteroid kill the dinosaurs?

About 65 million years ago, a major disaster happened on Earth. The dinosaurs, the most fearsome animals ever to roam our planet, mysteriously died out, and so did many other creatures. Scientists have found a strange layer of clay among the Earth's rocks that was formed at the time the dinosaurs died. The clay is unlike any other found on Earth. Scientists think that this clay is made up of dust from a large asteroid that hit the Earth in the Yucatán peninsula of Mexico. The dust would have spread around the globe and changed the Earth's climate for years before finally falling to the ground. The dinosaurs could not survive in this new climate, and died out. If another large asteroid hit the Earth today, it might be humans that die out.

CERES DATA

Distance from Sun:	260 million mi
Diameter:	580 mi
Time to orbit Sun:	4.6 years
Time to spin on axis:	9 hr

Meteors and meteorites

On any clear night you may see a sudden streak of light, often called a *falling star*. Actually, falling stars have nothing to do with stars at all. They are caused by specks of dust burning up as they plunge at high speed into the Earth's atmosphere. The proper name is a *meteor*.

Strung out along the orbits of comets are swarms of dust particles. Several times a year, the Earth passes through some of this dust. When this happens, a *meteor shower* is seen. During such a shower, dozens of meteors can be seen each hour, radiating from one part of the sky. A meteor shower can last several nights.

Occasionally, a far larger piece of debris enters the atmosphere and reaches the ground. This is known as a *meteorite*. Although the name is similar, meteorites are quite different from meteors. Meteorites are chips off asteroids. They can be of rock or metal.

The atmosphere slows down small meteorites so that they fall harmlessly to Earth, but the largest ones are still moving very quickly when they hit the ground so that they blast out a crater. Most of the craters on the Moon were caused by impacts from meteorites and asteroids. Scientists collect meteorites because they are free samples from other parts of the Solar System.

▲ Roughly 50,000 years ago a meteorite weighing thousands of tons crashed into the Arizona desert in the United States. The hole it made is known as Meteor Crater (although really it should be Meteorite Crater), 1200 meters (4000 feet) across and 180 meters (600 feet) deep.

▲ All meteors belonging to a particular meteor shower seem to come from the same point in the sky. This point is called the radiant of the shower.

The world's largest known meteorite is made of iron and weighs about 60 tons. It fell in prehistoric times in Namibia, Africa, where it still lies.

▲ A bright Perseid meteor streaks across the night sky. The Perseid meteor shower comes every August. It is named after the constellation Perseus, in which its radiant lies.

Stars and constellations

The stars may seem uncountable, but no more than about 2000 of them are visible to the naked eye on any clear night. In a town, you will be able to see just a few dozen of the brightest ones because of the glare from street lights. It is only when you look at the sky through binoculars or a telescope that the stars really do become impossible to count.

Stars seem to form patterns in the sky, called *constellations*. The Greeks and Romans of old named these star patterns after characters from their myths, which made the constellations easier to remember. We still use the old names today, as shown on the maps on the following eight pages.

More recently, astronomers have invented newer constellations for the stars in the southern skies. The Greeks and Romans could not see these stars because they never rose above the horizon in Greece and Italy. There are now 88 constellations, covering the whole sky. Although the stars of each constellation lie in the same direction in space, most of them are at vastly different distances from us. So there is no real connection between most of the stars in a constellation at all.

Star distances are often measured in *light years*, which is how far a beam of light travels in one year. Light travels at the fastest speed known, 300,000 kilometers (186,000 miles) per second. Therefore it takes light just over a second to reach us from the Moon, over 8 minutes to reach us from the Sun, and 4.3 years to reach us from the nearest star to the Sun, which is called Alpha Centauri and lies in the southern half of the sky. Many of the stars we see at night lie hundreds of light years away – in other words, their light has taken centuries to reach us.

How bright a star seems depends on both its actual brightness and how far away it is. For example, Sirius is the brightest star in the night sky. It gives out 20 times as much light as the Sun, but that alone does not account for its brightness at night. What is also important is that it is among the nearest stars to us, only 8.7 light years away. Some stars give out thousands of times more light than Sirius, but to us they appear much fainter because they are much farther away.

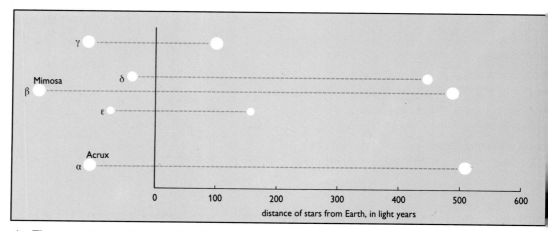

▲ *The constellation Crux, the Southern Cross, is shown on the left as it appears to us. If we could travel hundreds of light years through space and look at these five stars from a different viewpoint, we would see that they lie at very different distances from Earth and do not really form a cross at all. If we lived on a planet orbiting a distant star, we would see different patterns in the stars.*

The smallest of the 88 constellations is also one of the most famous — Crux, the Southern Cross. The ancient Greeks saw these stars as part of the legs of Centaurus, the centaur.

The changing shapes of the constellations

Stars are all moving through space at high speeds, but because they are so far away we do not notice any movement, even in a human lifetime. But over very long periods of time the shapes of the constellations will change. This diagram shows how the seven stars of the Big Dipper have moved over the past 100,000 years, and how they will look 100,000 years from now.

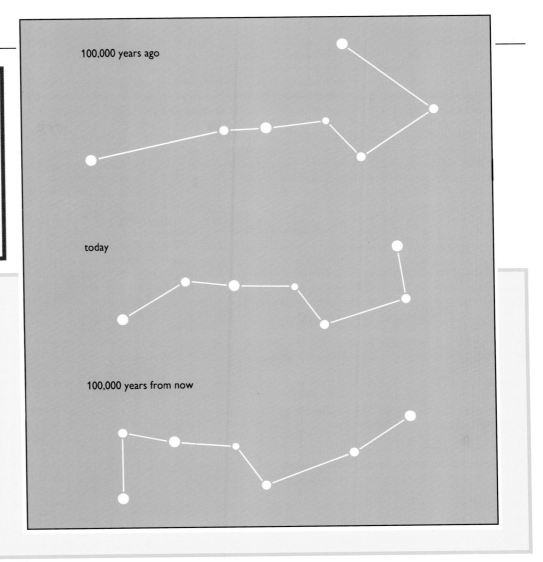

100,000 years ago

today

100,000 years from now

HÆMISPHÆRIUM AUSTRALE.   STELLATUM ANTIQUUM.

◀ *Maps of the night sky were once like this. As well as the stars themselves, they showed in great detail the mythical people, creatures, and objects after whom the ancient Greeks and Romans named the constellations. These two maps were made by the Dutch map-maker Andreas Cellarius in 1660. On the left are the constellations of the northern skies. The southern constellations are shown on the right-hand map.*

If you compare these two old maps with the star charts on the following pages, you will find that the constellations are not all the same. Some we no longer use. For example, Cellarius drew Cancer Minor, the little crab, beside Cancer itself, but there is no such constellation now. Around the south celestial pole are some gaps which were filled in with new constellations about a hundred years after these maps were made. The large constellation Argo, the ship, was later split into three: Carina, the keel, Puppis, the deck, and Vela, the sails.

Star charts

Between them, the star charts on the following pages cover the whole of the sky. The stars in the region around the north celestial pole are shown on these two pages. On pages 54–57 are charts of the sky either side of the celestial equator, and on pages 58–59 are the south polar stars.

When you look up at the night sky, the stars all seem to be on the inside of a huge dome. Although this isn't so, it gives us a way of mapping the stars and describing their positions. Maps of the Earth have a grid made up of lines of latitude and longitude. Instead, maps of the stars use lines of declination and right ascension. (These are marked on the charts as black dotted lines.) *Declination* is how far up or down a star is from the celestial equator. It is measured in degrees, from 0° at the equator to +90° (at the north celestial pole) or −90° (at the south celestial pole). *Right ascension* is how far around the star is, and is measured from 0 to 24 hours. You will find these coordinates marked around the edges of the charts.

The brighter stars are shown on the charts as bigger spots. Astronomers use a scale of *magnitudes* to measure star brightnesses, and the key by each chart gives the spot sizes for magnitudes of 5.5 and brighter. Stars as faint as magnitude 6 and even 6.5 can be seen by people with good eyesight in very clear, dark skies, but they are not on these charts. A small number of other objects are shown, like the globular cluster called 47 Tucanae, in the southern constellation Tucana (see page 67).

A white dotted line winding across the four equatorial maps marks the *ecliptic*. This is the path the Sun appears to follow across the sky as the Earth goes around it. The planets follow the ecliptic quite closely as they too orbit the Sun. The Milky Way is shown

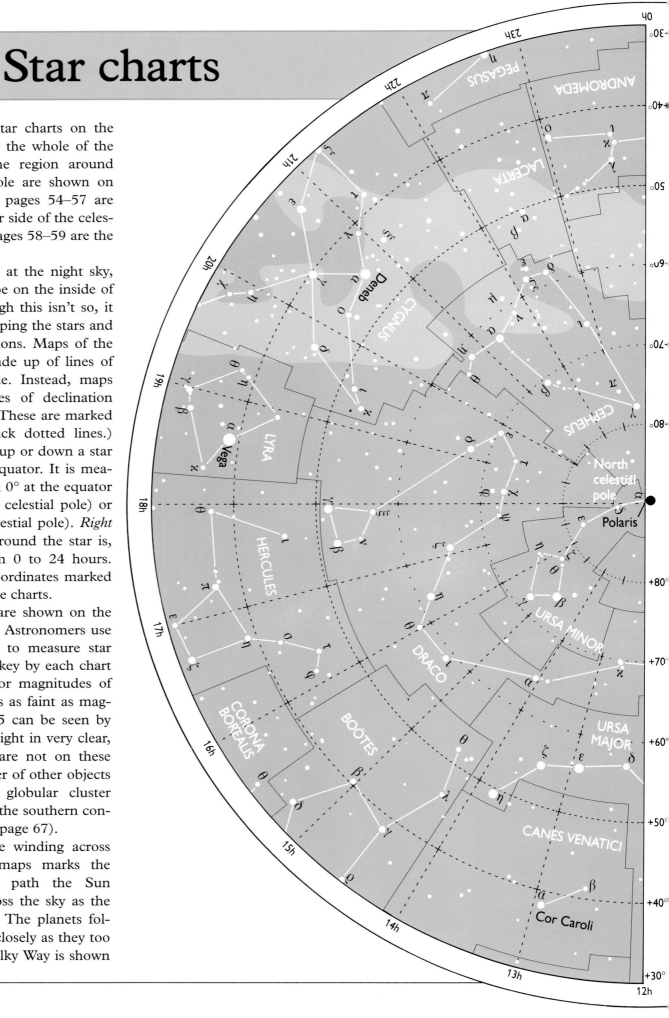

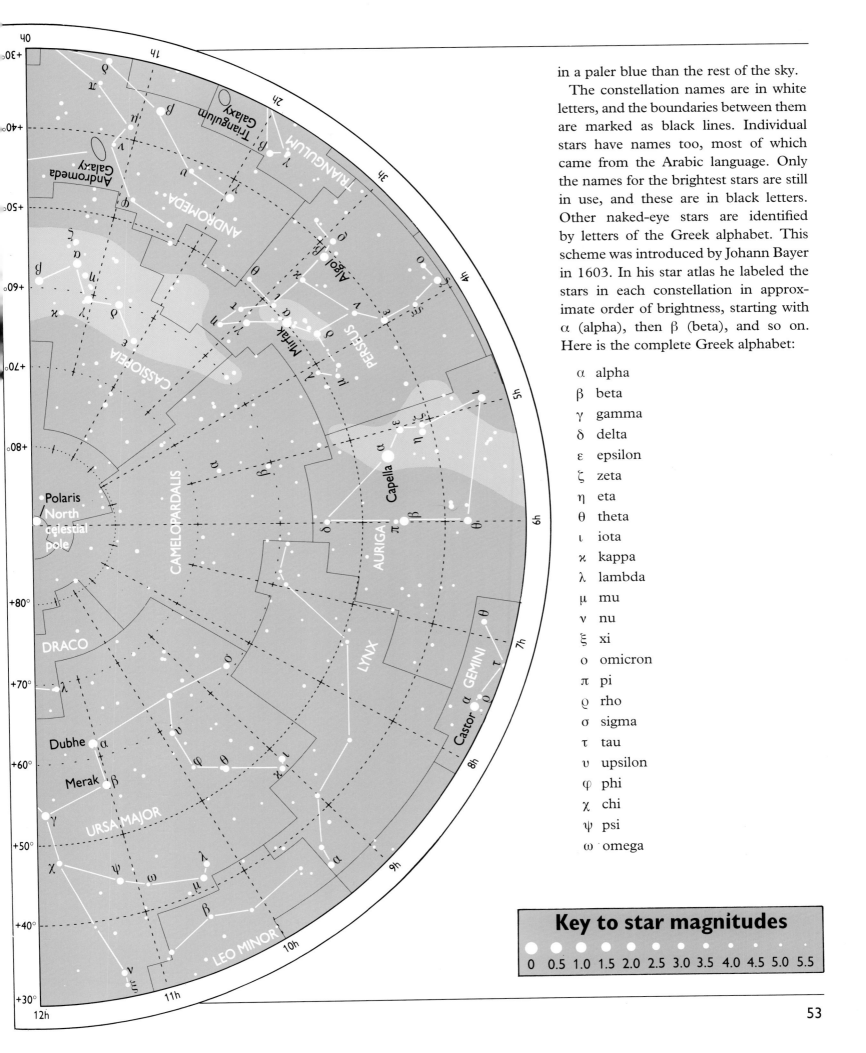

in a paler blue than the rest of the sky.

The constellation names are in white letters, and the boundaries between them are marked as black lines. Individual stars have names too, most of which came from the Arabic language. Only the names for the brightest stars are still in use, and these are in black letters. Other naked-eye stars are identified by letters of the Greek alphabet. This scheme was introduced by Johann Bayer in 1603. In his star atlas he labeled the stars in each constellation in approximate order of brightness, starting with α (alpha), then β (beta), and so on. Here is the complete Greek alphabet:

α	alpha
β	beta
γ	gamma
δ	delta
ε	epsilon
ζ	zeta
η	eta
θ	theta
ι	iota
κ	kappa
λ	lambda
μ	mu
ν	nu
ξ	xi
o	omicron
π	pi
ϱ	rho
σ	sigma
τ	tau
υ	upsilon
φ	phi
χ	chi
ψ	psi
ω	omega

Key to star magnitudes

0 0.5 1.0 1.5 2.0 2.5 3.0 3.5 4.0 4.5 5.0 5.5

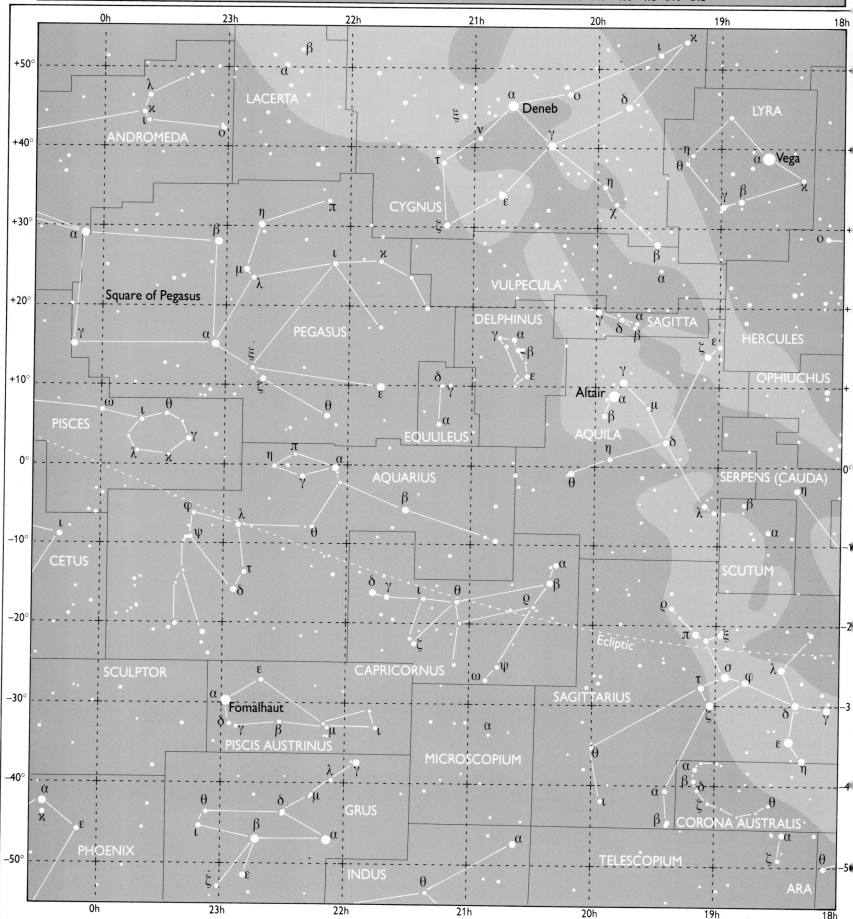

Key to star magnitudes

0 0.5 1.0 1.5 2.0 2.5 3.0 3.5 4.0 4.5 5.0 5.5

LACERTA

ANDROMEDA

β α LACERTA

λ χ ι o ANDROMEDA

ι χ LYRA

Deneb

α o δ LYRA

ξ ι

ν α Vega

τ CYGNUS η θ γ β χ

η ε γ β κ

π ζ χ

η β

μ λ ι χ β VULPECULA SAGITTA

Square of Pegasus γ α α

ξ PEGASUS DELPHINUS δ β HERCULES

γ α α γ α ζ ε OPHIUCHUS

ζ δ β

ω θ γ β ε Altair

PISCES ι θ δ γ α μ

λ γ χ γ AQUILA δ

CETUS η π α η

η γ EQUULEUS θ SERPENS (CAUDA)

β η

φ λ α λ β

ψ θ α SCUTUM

τ δ ρ α

δ γ ι θ β ρ

SCULPTOR ζ π ξ

CAPRICORNUS Ecliptic σ φ λ

ε ψ τ δ γ

α Fomalhaut ω SAGITTARIUS ζ ε

δ γ β μ ι α θ

PISCIS AUSTRINUS MICROSCOPIUM α β δ θ

λ γ ι α β ζ CORONA AUSTRALIS

θ δ μ α η

PHOENIX α ι β GRUS ι β α

χ ε β α ζ

PHOENIX θ ζ

ζ ε INDUS TELESCOPIUM θ ARA

DRACO

URSA MAJOR

BOÖTES

CANES VENATICI

Cor Caroli

CORONA
BOREALIS

COMA
BERENICES

LEO

SERPENS
(CAPUT)

HERCULES

Arcturus

Albireo

OPHIUCHUS

VIRGO

SERPENS
(CAUDA)

LIBRA

Spica

CORVUS

Ecliptic

CRATER

HYDRA

Antares

SCORPIUS

LUPUS

λ Shaula

CENTAURUS

o ω Centauri

ARA

NORMA

55

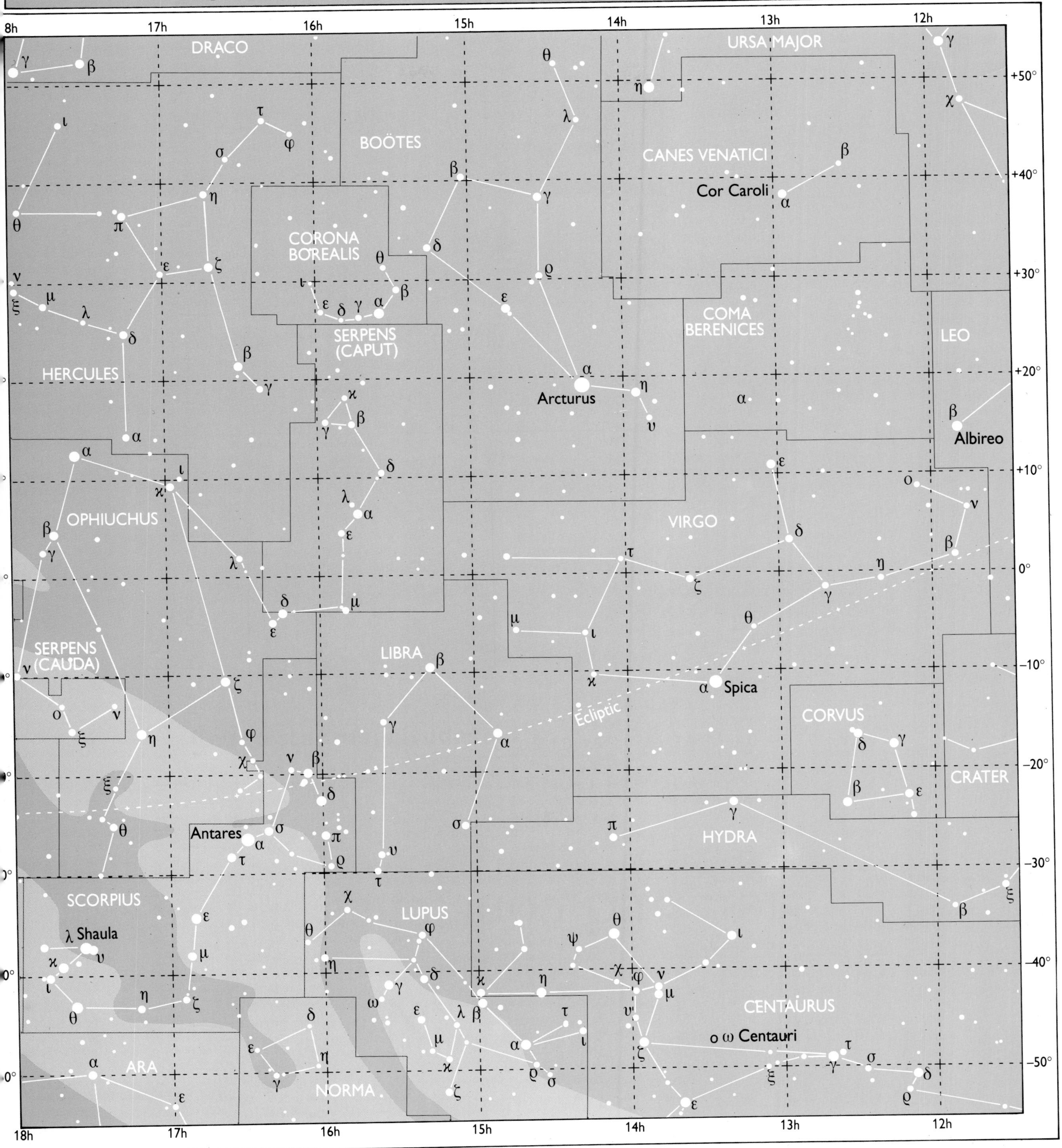

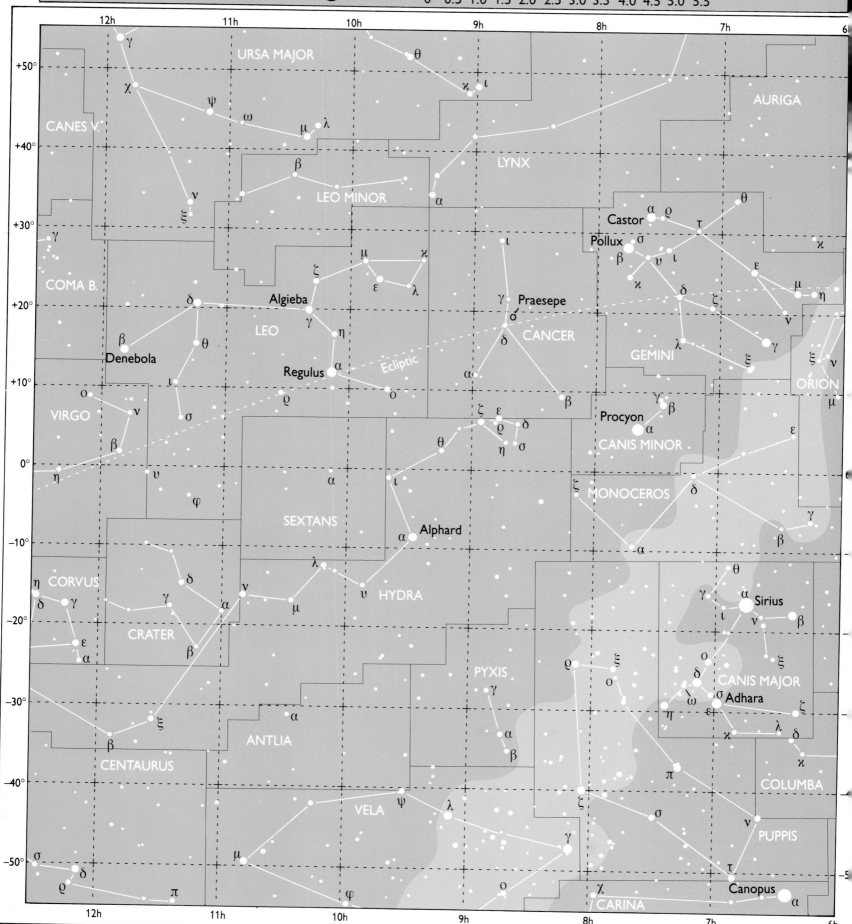

Key to star magnitudes

0 0.5 1.0 1.5 2.0 2.5 3.0 3.5 4.0 4.5 5.0 5.5

CASSIOPEIA

AURIGA

Capella

δ
π
β
θ

η
ε
ζ
ι

PERSEUS

Mirfak

Algol

TAURUS

Elnath

β

ζ

Pleiades

Aldebaran

Hyades

ORION

ANDROMEDA

Andromeda Galaxy

Triangulum Galaxy

TRIANGULUM

ARIES

PEGASUS

Ecliptic

PISCES

λ
α
Bellatrix
γ

Betelgeuse

ε
δ
ζ
σ
Orion
Nebula
ι

Rigel

χ

η
ζ
α
μ

δ
γ
β
ε

LEPUS

γ
α
ε
β

η

COLUMBA

PICTOR

β

γ

DORADO

CAELUM

α
α

HOROLOGIUM

γ

φ
χ

ERIDANUS

δ
ε
ζ
η

γ
π

α

FORNAX

θ
ι

χ

δ

Mira

o

CETUS

ζ
η

θ

τ

β

υ

ι

SCULPTOR

PHOENIX

γ
β

α
χ
ε

δ

57

Constellation index

This index gives the page number of the main star chart(s) on which to find each of the 88 constellations.

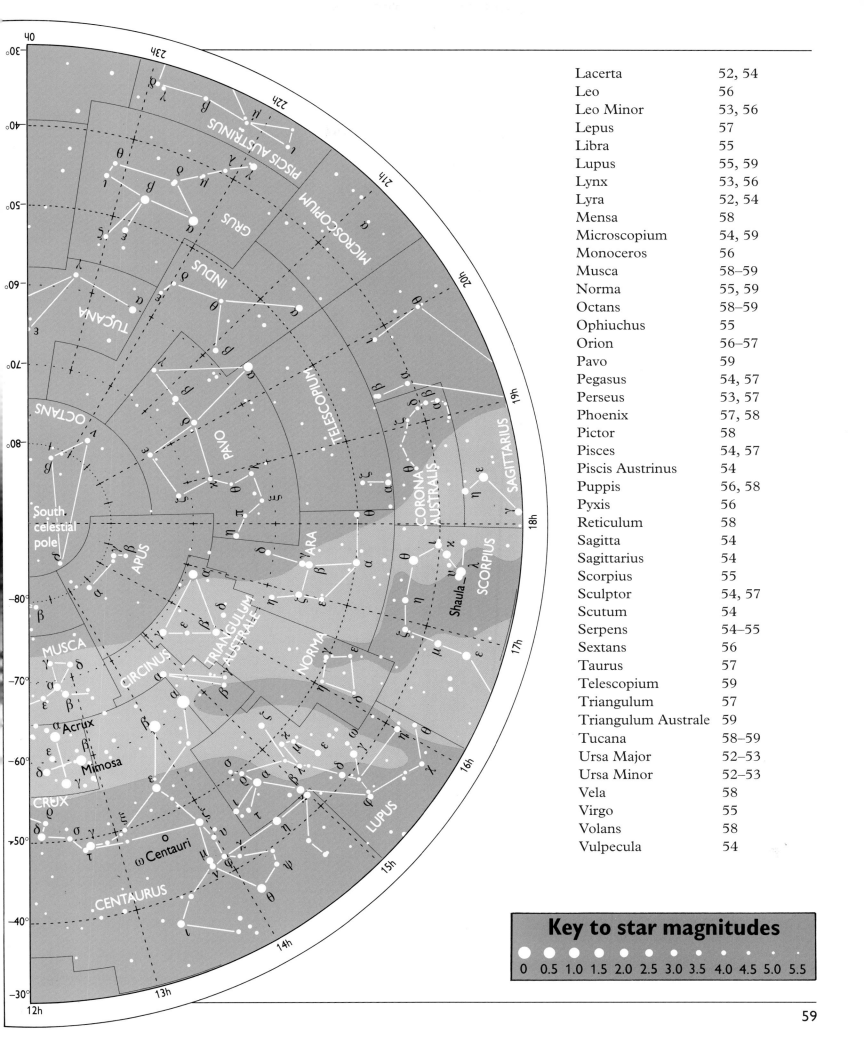

Key to star magnitudes

0   0.5   1.0   1.5   2.0   2.5   3.0   3.5   4.0   4.5   5.0   5.5

Finding your way by the stars

Bright stars and easy-to-spot constellations provide skymarks to help you find your way around the heavens. Find these stars and constellations on the maps on pages 52–59, and use them as signposts to other constellations nearby.

The stars on display at night will change from season to season as the Earth orbits the Sun. Also, which stars you can see depends on your latitude on Earth. And, of course, the night sky appears to spin around as the Earth rotates beneath it.

Finding north

The most familiar star group of all is the saucepan-shaped Big Dipper or Plough, actually part of the constellation Ursa Major (the great bear). Two stars of the saucepan's bowl point towards the north pole star, Polaris. In the opposite direction, these stars point to the constellation Leo, the lion. Following the handle of the saucepan takes you to Arcturus, the fourth-brightest star in the sky. If you have good eyesight, you will see that the second star in the handle of the saucepan has a twin. Binoculars will show the two stars more clearly. The best time of year to see the stars in this diagram is in the spring.

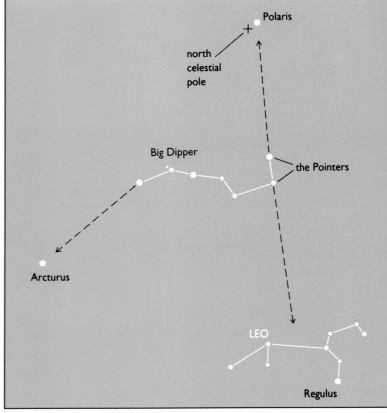

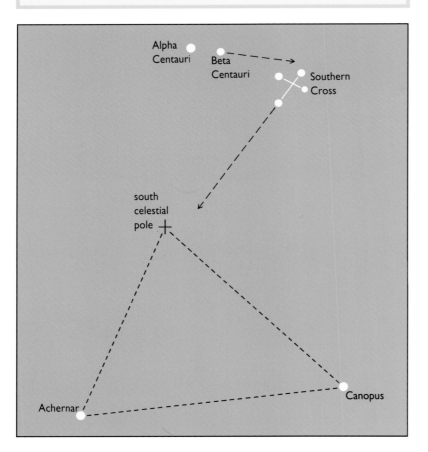

Finding the south pole

There is no bright pole star in the southern hemisphere. The Southern Cross, the constellation Crux, can be visualized as an arrow pointing to the south pole. The south pole of the sky forms a triangle with the bright stars Achernar and Canopus. Two other bright stars quite close together, Alpha and Beta Centauri, point to the Southern Cross. Alpha Centauri is the closest star to the Sun.

Sailors long ago used the stars to navigate at sea, as shown on this stamp from the Faroe Islands. Far away from the shore, with no landmarks to guide them, sailors took measurements of the stars at night to work out their position at sea. Great voyages of discovery were made possible by sailors' knowledge of the stars.

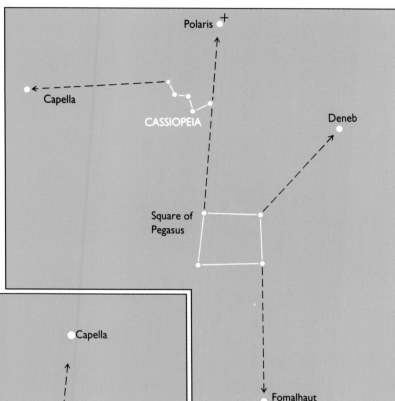

Stars in summer

In the northern hemisphere summer, three bright stars form a large triangle. They are called Deneb, Vega, and Altair. Vega is the brightest star in the constellation Lyra, the lyre. Deneb marks the tail of Cygnus, the swan. But the stars of the swan can just as easily be seen as a large cross. Sometimes, therefore, Cygnus is known as the Northern Cross.

Stars in winter

In the south on winter evenings lies the magnificent constellation Orion, the hunter. The star at its top left, Betelgeuse, forms a large triangle with Sirius (the brightest star in the sky) and Procyon. A line of three fairly bright stars forms the belt of Orion. Below the belt is the famous Orion Nebula, a glowing cloud of gas just visible to the naked eye in dark skies. Above and to the left of Betelgeuse are two stars representing the celestial twins, Castor and Pollux. To the top right of Orion is Aldebaran, the brightest star of Taurus, the bull.

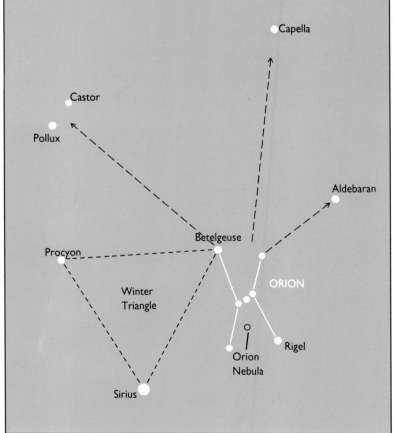

Stars in autumn

In the autumn sky lies the constellation Pegasus. Its main feature is a box of four stars, the Square of Pegasus. The sides of the square can be used to locate Polaris, the north pole star, and a bright star in the south called Fomalhaut. Between Pegasus and Polaris is the constellation Cassiopeia. This is easy to recognize, because it is shaped like the letter W.

Stars and star-birth

Space is mostly empty, but in places between the stars there are clouds of gas and dust called *nebulae*. Stars are born when gravity pulls the densest parts of a nebula into balls of gas. As the gas ball grows it gets so hot at its center that nuclear reactions start (see pages 6–7). These reactions give out light and heat, which is what makes the gas ball a true star. This is how the Sun was born 4.6 billion years ago. The planets grew from dust and gas that was left in orbit around it.

The energy-giving reactions inside a star turn hydrogen into helium. But when the hydrogen fuel in its core begins to run out, the star swells up in size. As the star swells its surface cools and turns redder. Such stars are called *red giants*. One day our Sun will become a red giant, a hundred times the size it is now. When that happens, about 5 billion years from now, the Earth will be roasted to a cinder.

◄ *An immense column of cool gas in the Eagle Nebula, photographed by the Hubble Space Telescope. New stars are forming from knots of denser gas at the top of the column.*

After this, the outer layers of the red giant drift off into space, like an enormous smoke ring. Such an object is called a *planetary nebula* – not because it has anything to do with planets, but because some astronomers thought they looked like the rounded disks of distant planets when seen through telescopes.

At the center of a planetary nebula lies the small, hot core of the former star. Such a core is known as a *white dwarf*. Over billions of years, the white dwarf cools and fades out. Our Sun will end up as a faint white dwarf.

▼ *Stars are being born in the Orion Nebula, a huge cloud of gas. This photograph, taken in infrared light, shows hundreds of young stars that have come into being from the nebula's gas within the past million years.*

White dwarfs

A white dwarf star is about the size of the Earth, but weighs as much as the Sun. Because so much matter is squeezed into such a small ball, white dwarfs are very dense. A spoonful of white dwarf material would weigh as much as a fully grown elephant. This stamp honors an Indian scientist, Subrahmanyan Chandrasekhar, who was awarded a Nobel prize for his studies of the strange nature of white dwarfs.

▼ *Four ages in the life of a star. On the left, collapsing gas in a nebula has become hot enough for nuclear reactions to begin, and a star starts to shine. Next, a few billion years later, it is shining steadily, as the Sun is now. Several billion years further on, it is a swollen red giant. Eventually (this part of the picture is on a much smaller scale), it slowly puffs away its outer layers into a planetary nebula.*

▲ *This artist's impression takes us 5 billion years into the future. The Sun is now a red giant, and we are looking at it from a burnt and lifeless Earth.*

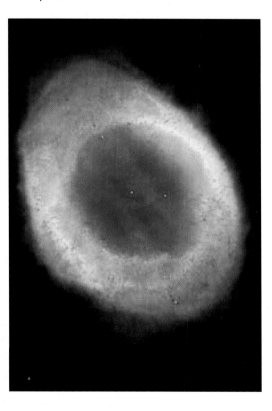

▲ *One of the most famous planetary nebulae is the Ring Nebula in the constellation Lyra. It consists of a cylinder of gas thrown off by a dying star, but it appears ring-shaped because we are looking down the length of the cylinder. The core of the former star is visible at the center.*

Supernova!

The brightest and heaviest stars suffer a much more spectacular death than ordinary stars like the Sun. After swelling up at the end of their life into brilliant supergiants they blow themselves apart in a huge explosion known as a supernova. A supernova shines as brightly as millions of ordinary stars like the Sun for a few weeks or months, before it fades away.

The outer layers of the star are blasted off into space at high speed. But at the center of the explosion something strange is left behind. In many cases, the core of the star is squashed by the supernova explosion to form a *neutron star*. Such stars are only about 20 kilometers (12 miles) across, the size of a city on Earth.

Because neutron stars are so small, they can spin very quickly – many times a second. Every time they spin they give out a flash of energy, like a lighthouse. Usually this energy is in the form of radio waves, but sometimes it can also

▲ *Perhaps this is what a neutron star would look like from close by.*

In 1572 a brilliant new star appeared in the constellation Cassiopeia, and was observed by the Danish astronomer Tycho Brahe. We now know that this star was a supernova. The remains of the exploded star can still be detected by astronomers. A plan of Tycho's observatory, one of his observing instruments, and the location of the supernova are shown on this stamp.

include light waves. A flashing neutron star is called a pulsar.

The matter in a neutron star is squeezed so incredibly tightly that a spoonful of it would weigh as much as a mountain. Its gravity is therefore very strong. But if a neutron star contains more matter than about three Suns, its gravity becomes so strong that it shrinks even further, until it vanishes from sight! Such an object is called a *black hole*. At the center of the black hole the star that died is crushed out of existence by the force of gravity.

Nothing can get out of a black hole, not even light. However, things can fall in, including gas from nearby stars, and this is how we can detect black holes. As gas swirls around the hole before falling out of sight, it heats up to many millions of degrees. Satellites in space have spotted a number of places where hot gas is believed to be falling into a black hole.

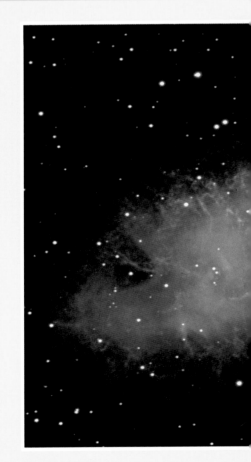

The Crab Nebula

In the year 1054, people on Earth saw a bright new star flare up for a few months in the constellation Taurus. We now know that this was a supernova. At the site of the explosion, the remains of the star that blew up can still be seen through telescopes: a vast expanse of glowing gas that is still expanding. It is called the Crab Nebula, because an astronomer last century thought it looked like a crab's claws. At the center of the Crab Nebula is a pulsar that flashes 30 times a second.

Special cameras that can take a thousand photographs a second have been used to catch the Crab Pulsar in the act of flashing. The sequence of nine small photographs shows the flash building up and then dying away.

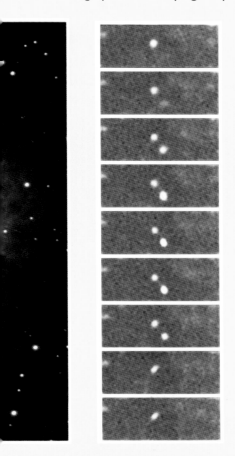

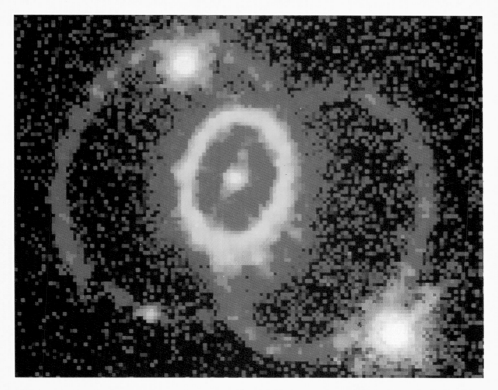

Supernova 1987A

In 1987, astronomers in the southern hemisphere were lucky to see the first naked-eye supernova for almost 400 years. Nearly all supernovae we know of have been recorded on photographs of far-distant galaxies. But Supernova 1987A exploded into view in a small galaxy next to ours called the Large Magellanic Cloud. Below are "before" and "after" photographs, an arrow pointing to the star before its supernova explosion. The photograph above, taken by the Hubble Space Telescope, shows rings of gas thrown off by the star before it exploded. The remains of the exploded star lie at the center of the middle ring.

Star families

Most stars are members of families – twins, triplets, or even larger groups. For example, through a telescope we can see that the closest star to the Sun, Alpha Centauri, is actually a pair of yellow stars. They orbit around each other every 80 years. Many other stars that look single to the naked eye turn out to have one or more partners when seen through telescopes.

In some pairs, the stars pass in front of one another as they go around their orbits, which causes an eclipse. When this happens, the light from one star is blocked off and we see the star fade for a few hours or days. The star returns to its normal brightness when the eclipse ends.

Young stars are often found in bunches of dozens or hundreds, all of them born from the same cloud of gas. One of the most famous star clusters is the Pleiades in the constellation Taurus, the bull. If you have good eye-sight you may be able to count six or more stars in the Pleiades. Binoculars reveal dozens of stars in the group. They all formed within the past 50 million years or so, which means they are young as stars go – the Pleiades did not exist when the dinosaurs were on Earth, for example.

Taurus contains another famous star cluster, the Hyades. This group, shaped like the letter V, is much older than the Pleiades. Its brightest stars can be seen with the naked eye and the rest are easily seen in binoculars.

Such groups are called *open clusters* because the stars are usually widely scattered. But there is another type of star group that contains hundreds of thousands of very old stars. These groups are rounded in shape and are called *globular clusters*. The stars in them are over twice as old as the Sun. They were among the first stars to have formed in our Galaxy.

▼ Seen through a telescope, Alpha Centauri (left) is a pair of yellowish suns. Epsilon Lyrae (middle) is a group of four stars – two are far apart, but each has a closer companion. Albireo (right), in the constellation Cygnus, is made up of a yellow and a blue star. The colors of the stars tell us what their temperatures are. The reddest stars are the coolest, and the bluest ones are the hottest.

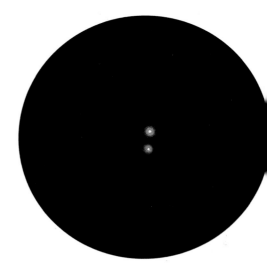

▼ The best-known star cluster, the Pleiades, is nearly 400 light years away. Its stars are immersed in a cloud of dust which gives the cluster a hazy appearance on this photograph.

Taurus, the bull, contains two famous clusters of stars. The V-shaped Hyades cluster forms the face of the bull, while the smaller Pleiades cluster lies on the bull's back. The brightest star in Taurus is Aldebaran, which marks one of the bull's eyes.

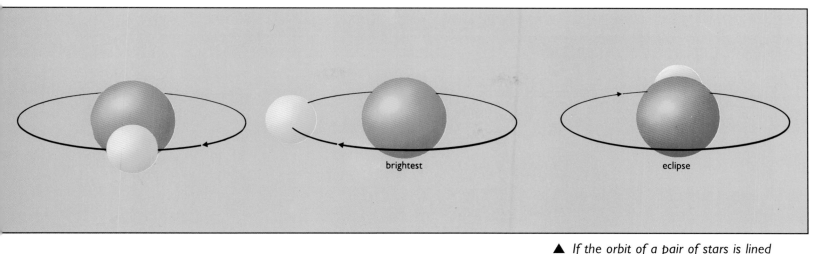

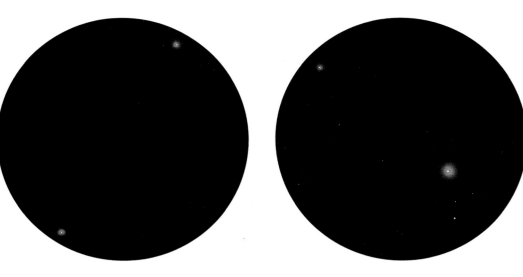

▲ If the orbit of a pair of stars is lined up in a particular way as seen from Earth, they will pass in front of each other and the brightness will appear to change, especially when a large, dim star passes in front of a much brighter companion. The most famous star of this type is Algol, in the constellation Perseus. Its name means "the winking demon."

▼ This globular cluster (below left) is known as 47 Tucanae. It contains something like 100,000 stars. The smaller photograph below was taken with the Hubble Space Telescope and shows individual stars in the dense central regions of another globular cluster, called M92.

Our Galaxy

globular clusters

nucleus

spiral arm

position of
the Sun

The Magellanic Clouds

Our Galaxy has two small neighbors. They both lie in the southern half of the sky and look like separate parts of the Milky Way. They are the Magellanic Clouds, named after the 16th-century Portuguese explorer Ferdinand Magellan who saw them on his voyages around the world. The Large Magellanic Cloud is about one-tenth the size of our Galaxy and lies about 170,000 light years away. The Small Magellanic Cloud is smaller and more distant. Both are irregular in shape. This photograph shows the Large Magellanic Cloud. The pink patches are nebulae (gas clouds) where stars are forming. The biggest such area, called the Tarantula Nebula because it looks like a large tarantula spider, is 30 times the size of the Orion Nebula in our Galaxy.

This artist's impression shows our Galaxy as it would look from the outside. The Sun lies in one of the arms that spiral out from the hub. (The Sun is shown larger than it really is, otherwise it would be invisible.) Around the center is a halo of globular clusters, colored orange in this picture.

Our Sun and the stars that make up the constellations are all part of a huge grouping called the Galaxy. From our position inside the Galaxy it is difficult to see what it looks like. But astronomers think that its shape is a huge spiral, rather like a coil of rope.

The Galaxy is about 100,000 light years wide, which means that a beam of light (or a radio signal) would take 100,000 years to cross it. Our Sun lies in one of the spiral-shaped arms of the Galaxy, about two-thirds of the way from the middle of the Galaxy to its rim, so we are in the suburbs. Beyond our Galaxy is empty space, and then other galaxies (our Galaxy is given a capital letter).

On a clear night you can see the distant stars in our Galaxy. They form a faint band of light across the sky, the Milky Way. If you look at the Milky Way with binoculars or a telescope, you will see that it is made up of stars crowded together in their thousands. Our Galaxy itself is sometimes called the Milky Way.

There are over 100 billion stars in the Galaxy. This is only a rough guess, for no one has counted them all. If you tried counting them at the rate of 100 a minute, without stopping, it would take you at least 2000 years!

As well as middle-aged stars like the Sun, the spiral arms of the Galaxy contain young stars in open clusters and gas clouds (nebulae) where more stars are still forming. Dotted around the Galaxy are globular clusters, full of old stars. Old stars are also found in the center of the Galaxy, which lies in the constellation Sagittarius.

The whole Galaxy is rotating. As with the planets of the Solar System, the stars nearest the center go around quickest and those farthest away take the longest. The Sun goes around the Galaxy in about 220 million years. It has completed only about 21 orbits since it was born.

The center of our Galaxy lies in the constellation Sagittarius, the Archer, where the stars of the Milky Way crowd together.

▼ *The beautiful Orion Nebula sits below the three stars of the "belt" in the constellation Orion, 1300 light years away in the same spiral arm of the Galaxy as our Sun.*

◄ *Towards the center of the Galaxy, the stars of the Milky Way seem to be packed so closely together that in photographs like this it is hard to see the individual stars. We would see even more of them were it not for the dark clouds of dust and gas that block part of our view.*

Galaxies and the Universe

Our Galaxy is one of countless galaxies dotted throughout the Universe, like islands in a vast ocean. Galaxies come in three main shapes. Many of them are *spirals*, like our own Galaxy. In spiral galaxies, stars and gas clouds lie in arms that curl out from the hub. Slightly different are the *barred spirals*. These have a bar of stars across their center, and the spiral arms start at the ends of the bar.

Totally different are *elliptical* galaxies. These have no arms at all – instead, they are rounded in shape. Elliptical galaxies come in a wide range of sizes. The largest ellipticals contain over ten times as many stars as the Milky Way and are the biggest galaxies known. But the smallest elliptical galaxies are just like large globular clusters.

Some galaxies have no particular shape at all, and are known as *irregulars*. The two Magellanic Clouds that accompany our Galaxy are examples.

Most galaxies belong to groups. Our own Galaxy is the second-largest member of the Local Group, which contains about three dozen galaxies. The largest member of the Local Group is a spiral galaxy in the constellation Andromeda. It can just be seen with the naked eye on clear nights as a fuzzy

▲ *The Andromeda spiral galaxy is very much like our own, but about twice as big. It too has tw[o] small galaxies as companions. They are visible in this photograph, one above it and one below.*

patch, and is easily found in binoculars.

The Andromeda spiral galaxy is a slightly larger version of our own Galaxy. It lies about 2.5 million light

▼ *A giant elliptical galaxy, M87, surrounded by hundreds of globular clusters.*

years away, which means that th[e] light from it now reaching us left the[re] while ape-men lived on Earth. Th[e] Andromeda spiral galaxy is the mo[st] distant object that we can see wi[th] the naked eye, without binoculars or [a] telescope.

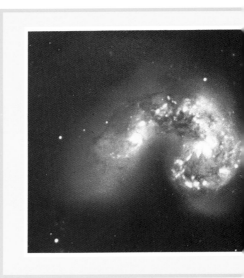

▲ A barred spiral galaxy, known as NGC 1365. (Most galaxies are known by numbers given to them in lists drawn up by astronomers of the past.) The arms are bluer than the bar because they contain younger stars.

The spiral galaxy NGC 2997, shown on an Australian stamp. The photograph used for the stamp was taken through the Anglo-Australian Telescope, which is in New South Wales, Australia.

Interacting galaxies

Most galaxies lie far apart, but sometimes one may pass close to another and may even collide with it. One famous example of a galaxy being disturbed by a passer-by is the Whirlpool (above), a beautiful spiral with a smaller companion. The smaller galaxy is actually in orbit around the Whirlpool, and the two probably brushed past each other millions of years ago. In another pair, called the Antennae (left), two spiral galaxies of similar size have passed close to each other, and gravity has pulled out two long streams of stars and gas, like the feelers of a giant insect (just visible at the top of the photograph). Most remarkable of all is Centaurus A (right), an elliptical galaxy with a band of dust around it. Astronomers think that Centaurus A was formed when an elliptical galaxy merged with a spiral galaxy.

The origin of the Universe

In the 1920s, an American astronomer named Edwin Hubble made a sensational discovery: the Universe is getting bigger. He found this out by studying galaxies far off in space through the 2.5-meter (100-inch) telescope on Mount Wilson in California. At that time it was the largest telescope in the world.

As Hubble looked farther and farther into the depths of space, he noticed that the galaxies appeared to be moving apart. In other words, the entire Universe is swelling up like a balloon.

This is an important clue to how the Universe began. Astronomers now

A lens in space

This picture shows four images of the same quasar, looking like a four-leaf clover. At the center is the fuzzy image of a galaxy that lies directly between us and the quasar. The gravity of the galaxy acts like a lens, bending the quasar's light around it and splitting it into four parts. This photograph was taken by the Hubble Space Telescope, which is named after the man who discovered that the Universe is expanding.

A Seyfert galaxy, a type of spiral with a brilliant center. A quasar might look like this if seen close-up.

Looking back in time

Large telescopes can pick out objects so far away that their light started out on its journey to us before the Earth was born. Therefore, by looking far enough out into space, we can see how the Universe appeared shortly after the Big Bang, billions of years ago. At that time, the Universe contained many brilliant objects called *quasars*. A quasar is hundreds of times brighter than our Galaxy, but only a few times the size of our Solar System. Quasars are thought to be young galaxies with huge black holes at their centers, swallowing gas and even complete stars. As the gas swirls around the black hole before plunging in, it gets hot and shines brightly. Similar events are happening in what are called Seyfert galaxies, which are spirals with very bright centers.

▲ The Big Bang started the Universe expanding. It still is expanding, and the millions and millions of galaxies are all moving away from one another. Astronomers are not sure whether the Universe will go on expanding for ever, or whether it will stop and begin to contract – if it does this, it may end with a "big crunch."

think that, long ago, the entire Universe was contained in a single, incredibly dense blob. For some unknown reason, the blob exploded. The explosion is known as the Big Bang. The galaxies are the bits from that explosion, still flying outwards.

From the speed at which the galaxies are moving, we can work out roughly how long ago the Big Bang happened. The answer is about 15 billion years, about three times the age of the Sun and Earth.

At the time of the Big Bang, the temperature of the Universe was many millions of degrees. It has since cooled to only 3°C (5°F) above the coldest temperature possible, absolute zero. Absolute zero is –273°C (–459°F), so the temperature of the Universe is –270°C (–454°F). There have been other theories of how the Universe came to be, but the Big Bang is the best one so far.

In 1965 two American astronomers, Arno Penzias and Robert Wilson, discovered that empty space is not entirely cold. It has a temperature of 3 degrees above absolute zero, due to heat left over from the Big Bang. In 1978 Penzias and Wilson received a Nobel prize for their discovery.

Telescopes and binoculars

Telescopes show objects that are too faint to be seen with the naked eye, and they make them appear closer. To do these two seemingly magical tricks, they collect light rays into an image (picture) of the object and then enlarge it.

One type of telescope has a lens at the front of its tube. This is known as a *refracting* telescope. The lens refracts (bends) the incoming light to form an image. A smaller lens, called the eyepiece, magnifies (enlarges) the image so that the object appears closer. For example, through a telescope with an eyepiece that magnifies 100 times, we would seem to hover less than 4000 kilometers (2400 miles) above the Moon's surface.

In *reflecting* telescopes, a curved mirror bounces light into the eyepiece via a smaller second mirror. Some telescopes use both lenses and mirrors, but it is always the size of the main lens or mirror that is most important. Larger lenses and mirrors show fainter objects and smaller details because they collect more light. The most powerful telescopes can see objects 10 million times too faint for the naked eye.

Big mirrors are cheaper and easier to make than big lenses, so all large telescopes are reflecting ones. Very large mirrors are now being made from many smaller pieces fitted together. Two major examples are the twin Keck Telescopes which have mirrors 10 meters (33 feet) wide made from 36 pieces of glass. They are on a high mountain called Mauna Kea in Hawaii, where the skies are particularly clear. Telescopes with single mirrors over 8 meters (26 feet) across are now being built.

The world's largest refractor, at Yerkes Observatory in the United States, has a lens 1 meter (40 inches) across and was built as long ago as 1897. It is still used for research.

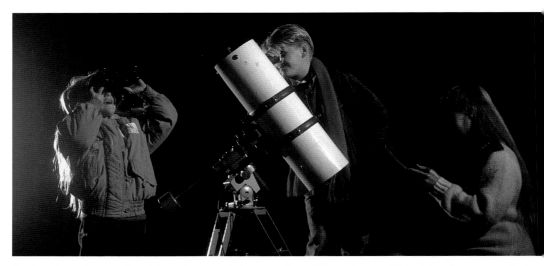

▲ *An observing session. The telescope being used is a 15-centimeter (6-inch) reflector, a popular choice for amateur astronomers. It is important to keep warm while looking at the night sky. Even though this is a summer's night, all three observers are well wrapped up.*

Binoculars

Binoculars are like two small telescopes joined together, so that you can look through them with both eyes. They are smaller and easier to carry than a telescope and are good for simple star-spotting. Light passing through binoculars is "folded" by two wedges of glass called prisms, which makes them shorter.

Binoculars carry markings such as 8 x 40 or 10 x 50. The first figure is the magnification, and the second figure is the size of the front lenses in millimeters. Binoculars usually magnify between 6 and 10

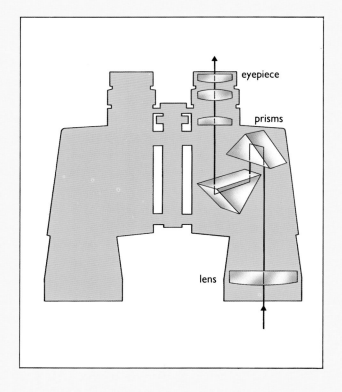

times and have front lenses from 30 mm to 50 mm (1 to 2in) wide. Binoculars give beautiful wide-angle views of the sky. They are useful for sweeping over the star fields of the Milky Way, looking at large star clusters, observing comets, and picking out stars and nebulae that are too faint for the naked eye.

How to observe

People are often disappointed with their first view through a telescope. But with care and practice, your observing will be rewarding. First, even if you are not using binoculars or a telescope, find somewhere safe to observe that is away from the glare of artificial lights. Most important is that your eyes will need 20 minutes or so to get used to seeing faint objects in the dark, so be patient. Astronomers need to "train" their eyes – the more you observe, the more you will see. And do keep notes of your observations: what you have seen, and the time.

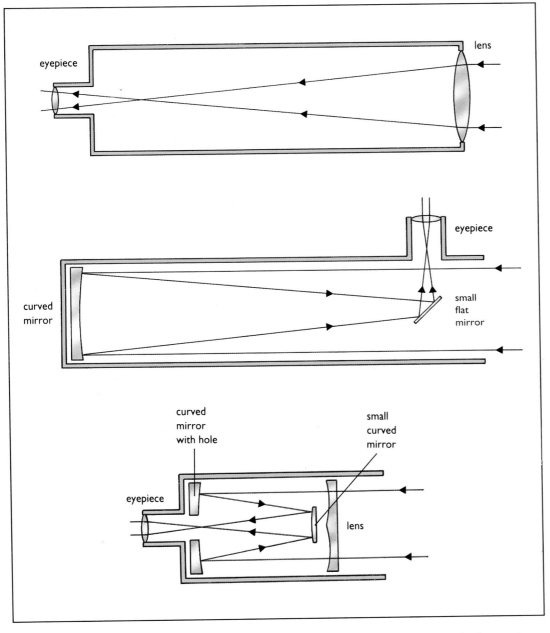

▲ These "cutaway" diagrams show how the three main types of telescope work. At the top is a refractor, which focuses light with a lens. The middle diagram shows a reflector, which uses a curved mirror instead of a lens. The telescope at the bottom uses mirrors and a lens to bring light to a focus. All telescopes have a lens for an eyepiece.

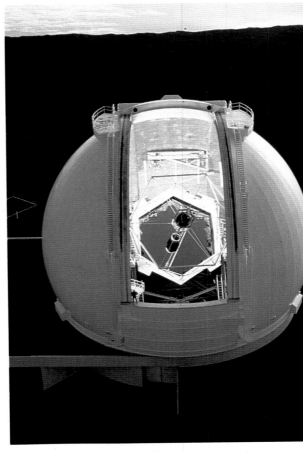

▲ One of the twin Keck Telescopes, which sit next to each other on a mountain in Hawaii. They each have mirrors 10 meters (33 feet) across. This view shows one of the telescopes through its open dome.

In 1990 the Hubble Space Telescope was launched into orbit by the Space Shuttle Discovery. It has a mirror 2.4 meters (94 inches) wide, but it can see the sky more clearly than smaller telescopes on the ground because it is above the Earth's atmosphere.

Seeing the invisible

▲ The Astro-1 telescopes aboard the orbiting Space Shuttle Columbia observed the sky at ultraviolet and x–ray wavelengths. The first flight of Astro 1 was in 1990.

▲ The spiral galaxy M81, photographed in ultraviolet light through a telescope called Astro-1 aboard the Space Shuttle. The bright blobs in the spiral arms are hot, young stars.

Objects in the Universe do not just give out light we can see. They also give out a whole range of other waves, from radio waves to X-rays, which we cannot see. However, these waves can be picked up by special instruments on Earth and in space.

Radio telescopes collect radio waves, and are usually in the form of large metal dishes, rather like reflecting telescopes. They are very much larger than optical telecopes because radio waves are much longer than light waves. The largest radio astronomy dish is 305 meters (1000 feet) wide, at Arecibo in Puerto Rico. Often, many smaller radio telescopes are joined together in a long line. This gives a more detailed picture of the sky than one single dish could produce alone.

Radio waves get though the atmosphere. Most other waves do not, and so they have to be studied from space. Different types of wave are given out by different types of object. For example, infrared rays are a good way of studying cool stars and clouds of gas in space. A satellite called IRAS (InfraRed Astronomical Satellite) studied star-birth and also found signs of planets forming around several nearby stars.

Very hot gas gives out the shortest wavelengths – ultraviolet, X-rays, and gamma rays. Some of the most exciting discoveries have come from X-ray satellites, which have found places in space where gas seems to be swirling around black holes.

Most of these astronomy satellites work by remote control from Earth but some telescopes have been operated by astronauts in space.

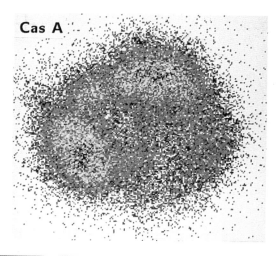

Cas A

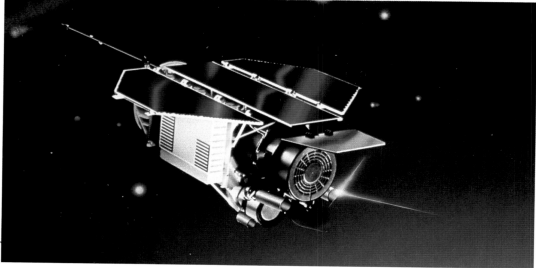

▲ The Very Large Array is a suitable name for this group of 27 radio telescopes in New Mexico. They can be moved along three sets of rails, each 21 kilometers (13 miles) long, in a Y shape. Each dish is 25 meters (82 feet) across. Here eight of them are lined up together.

◄ ROSAT took this x–ray picture of an object called Cassiopeia A (Cas A for short), which is also a strong source of radio waves. Astronomers think it is an expanding cloud of dust thrown out by a supernova explosion in 1667, although there is no record of anyone seeing a supernova that year.

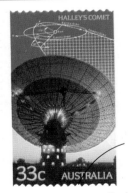

The radio telescope at Parkes, Australia, has a dish 64 meters (210 feet) in diameter. It has been used to study quasars and pulsars, and this stamp commemorates its role in observing Halley's Comet.

◄ This is an artist's impression of ROSAT, a German satellite for looking at the sky in X-rays, launched in 1990.

▲ An artist's impression of IRAS, the Infrared Astronomical Satellite, in orbit. During ten months in 1983 this telescope looked at a quarter of a million objects that send out infrared radiation, including stars, galaxies and dust clouds. It even discovered some comets.

The spectrum

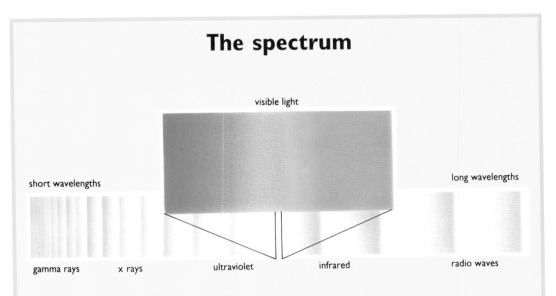

visible light

short wavelengths

long wavelengths

gamma rays   x rays   ultraviolet   infrared   radio waves

Light has a range of colors from blue to red, as in a rainbow. This band of colors is known as the *spectrum*. The color of light depends on the length of the waves that make it up. Blue light has the shortest wavelength and red light the longest. Beyond the blue and red ends of the spectrum are other wavelengths we cannot see. Ultraviolet waves are shorter than blue light. Shorter still are x rays and gamma rays. These short wavelengths can be detected only by instruments in space, since they do not get through the Earth's atmosphere. At the other end of the spectrum, infrared rays are longer than red light. Longest of all are radio waves, which are picked up by radio telescopes on Earth. All these waves travel at the same speed – the speed of light, 300,000 kilometers (186,000 miles) per second, the fastest speed in the Universe.

Quiz

1. How many planets are there in the Solar System?
2. Which is the biggest planet in the Solar System?
3. Which is the smallest planet in the Solar System?
4. Which planet comes closest to Earth?
5. How many planets lie between the Earth and the Sun?
6. How many miles away from us is the Sun, roughly?
7. How old are the Sun and the Earth, roughly?
8. Who was the first man to walk on the Moon?
9. Which planet is known as the red planet?
10. Which planet has a Great Red Spot?
11. Where in the Solar System would you find Maxwell Montes?
12. Where in the Solar System would you find Olympus Mons?
13. Where in the Solar System would you find the Ocean of Storms?
14. What's the name of the European space probe that photographed Halley's Comet in 1986?
15. Which space probe flew past the planets Jupiter, Saturn, Uranus, and Neptune?
16. Which is the only moon in the Solar System with thick clouds?
17. What's the name of Jupiter's volcanically active moon?
18. Which is the largest moon in the Solar System?
19. What are the names of the two small moons of Mars?
20. Who discovered the planet Uranus?
21. What's the closest star to the Sun?
22. How many light years away is the closest star to the Sun, roughly?
23. How many constellations are there?
24. Which constellation is pictured as a bull?
25. Which constellation is pictured as a lion?
26. Which constellation represents a swan?
27. What's the brightest star in the night sky?
28. What is the name of the north pole star?
29. What's the name given to the explosion of a big star at the end of its life?
30. What type of star will be left when the Sun dies?
31. What's the distance from one side of our Galaxy to the other in light years, roughly?
32. What shape is our Galaxy?
33. What are the two small companions of our Galaxy called?
34. Who discovered that the Universe is expanding?
35. What's the speed of light in miles per second, roughly?
36. Are the largest telescopes reflectors or refractors?

Answers on page 80.

Index

Quiz answers

1. Nine.
2. Jupiter.
3. Pluto.
4. Venus.
5. Two.
6. 93 million.
7. 4.6 billion years.
8. Neil Armstrong.
9. Mars.
10. Jupiter.
11. Venus.
12. Mars.
13. The Moon.
14. *Giotto.*
15. *Voyager 2.*
16. Titan.
17. Io.
18. Ganymede.
19. Phobos and Deimos.
20. William Herschel.
21. Alpha Centauri.
22. 4.3 light years.
23. 88.
24. Taurus.
25. Leo.
26. Cygnus.
27. Sirius.
28. Polaris.
29. Supernova.
30. White dwarf.
31. 100,000.
32. Spiral.
33. The Magellanic Clouds.
34. Edwin Hubble.
35. 186,000.
36. Reflectors.